I0126971

Books by James DeMeo

Saharasia: The 4000 BCE Origins of Child Abuse, Sex-Repression, Warfare and Social Violence In the Deserts of the Old World, Revised Second Edition 2006.

The Orgone Accumulator Handbook: Wilhelm Reich's Life-Energy Discoveries and Healing Tools for the 21st Century, with Construction Plans, Third Revised Edition 2010.

Preliminary Analysis of Changes in Kansas Weather Coincidental to Experimental Operations with a Reich Cloudbuster: From a 1979 Research Project, Orgone Biophysical Research Lab, Ashland, Oregon, 2010.

(as Editor) *Heretic's Notebook: Emotions, Protocells, Ether-Drift and Cosmic Life-Energy, with New Research Supporting Wilhelm Reich,* Orgone Biophysical Research Lab, Ashland, Oregon, 2002.

(as Editor) *On Wilhelm Reich and Orgonomy,* Orgone Biophysical Research Lab, Ashland, Oregon, 1993.

(as Co-Editor) *Nach Reich: Neue Forschungen zur Orgonomie: Sexualökonomie, Die Entdeckung der Orgonenergie,* Zweitausendeins Verlag, Frankfurt, 1997.

For a full list of James DeMeo's publications, see:
www.orgonelab.org/demeopubs.htm

In Defense of
Wilhelm Reich

In Defense of Wilhelm Reich

Opposing the *80-Years' War* of
Mainstream Defamatory Slander Against
One of the 20th Century's Most Brilliant
Physicians and Natural Scientists

by

James DeMeo, PhD

Natural Energy Works
Ashland, Oregon, USA
www.naturalenergyworks.net

Publication and worldwide distribution rights:

Natural Energy Works
PO Box 1148
Ashland, Oregon 97520
United States of America
http://www.naturalenergyworks.net

Email: info@naturalenergyworks.net

Also available through Lightning Source / Ingram Distribution

IN DEFENSE OF WILHELM REICH, Copyright © 2013 by James DeMeo. All Rights Reserved. Printed in the United States of America. No part of this book, text or images, may be used or reproduced in any manner without written permission from the author, except in the case of "fair use" quotations embodied in critical articles or reviews, with proper citation. For information, send inquiries to Natural Energy Works.

ISBN: 978-0-9802316-7-0, 0-9802316-7-1

First Edition, 2013

131222

Cover photo of Wilhelm Reich, courtesy of Renata Moise. Reich photo on p.xvi adapted from Ellen Siersted's *Wilhelm Reich in Denmark*. Reich photo on p.76 courtesy of the *Wilhelm Reich Museum and Trust*. http://www.wilhelmreichtrust.org

Shallow ideas can be assimilated. Ideas that require people to reorganize their picture of the world provoke hostility.
> \- Leo Tolstoy

We are accustomed to having men jeer at what they do not understand.
> \- Goethe

Unless we put medical freedom into the Constitution, the time will come when medicine will organize into an undercover dictatorship... The Constitution of this republic should make special privilege for medical freedom as well as religious freedom.
> \- Benjamin Rush
> 1776, Signer of the
> Declaration of Independence

When a true genius appears you can tell him by this sign, that all the dunces are in a confederacy against him.
> \- Jonathan Swift

In Defense of Wilhelm Reich

CONTENTS

Page

Preface and Acknowledgments.. xiii

1. Introduction   1

2. Wilhelm Reich Timeline:
    Biographical Summary, Discoveries and Attacks   13
    Reich's European Period   13
    Reich's American Period   48

3. After Reich: Verification of Reich's Findings by Other
    Scientists and Physicians, with Renewed Assaults ...   89
    Attacks on Reich's Research Legacy Postmortem   89
    Personal Experiences   95
    Recent Independent Confirmations of
        Reich's Discoveries.. 100

4. Critical Review of Christopher Turner's
    Adventures in the Orgasmatron 115
    Turner's 2004 Swipe at Reich, Orgonomy,
        and Summerhill School 118
    Turner's 2011 "Orgasmatron" book 120
    Turner's Background and Credentials 121
    Repeating Old Slanders from Brady and Gardner 122
    Reich and Einstein, Misrepresented 127
    Influencing America: Reich Versus Kinsey 130
    Turner's Serious Omissions on Kinsey 132
    Reich Versus Kinsey on Orgastic Functions 141
    Reich and the "Bohemians" 144
    More Nasty Accusations: Madness and Porn 146
    American Conditions and Sexual Freedom 151
    Reich's Bioelectric Experiments, and the Lie
        About "Masturbation of Patients" 155
    Silvert's Betrayal of Reich 160

CONTENTS (Continued)

4. (Continued) Critical Review of Christopher Turner's
   Adventures in the Orgasmatron

    Aurora Karrer's Confusions, and
        Reich's Last Years 166
    The Death of Troll 169
    Summary 172
    Open Letter to Jonathan Galassi, head of
        Farrar, Straus and Giroux 176

5. The Orgasmatron Aftermath: Defamatory Slanders
    From UK Channel 4 / Wag-TV 179
    From *Nature* Magazine 188
    From Wikipedia 197

6. Dusan Makavejev's Porno Film:
    WR Mysteries of the Organism 201

7. Orgone Flapdoodle:
    Enthusiast-Mystic Internet Nonsense 211

Appendix
    CSICOP, Prometheus Books, Pornography
        and the *Journal of Pedophilia?!* 219
    Related Items to Consult on the "Skeptics" 222

References and Notes 223
    Additional Materials for Consultation 252
    Major Works by Wilhelm Reich 254

Index 255

About the Author 269

Abbreviations and Unusual Terms

ACLU	American Civil Liberties Union
AMA	American Medical Association
APÖ	(SDAP) Austrian Social Democrat Worker's Party
ATIC	Air Technical Intelligence Command
Comintern	Communist International, Moscow controlled
CU	Consumers' Union
DOR	Deadly Orgone Radiation
DPG	German Psychoanalytic Society
FBI	Federal Bureau of Investigation
FDA	Food and Drug Administration
HUAC	House Committee on Un-American Activities
IJSO	*Int. Journal for Sex-Economy & Orgone Research*
INS	Immigration and Naturalization Service
IPA	International Psychoanalytic Association
ISU	Illinois State University
IZP	*Int. Zeitschrift (Journal) für Psychoanalyse*
KGB	Soviet Secret Service (Post-NKVD)
KPD	Communist Party of Germany
KPÖ	Communist Party of Austria
LRB	*London Review of Books*
KU	University of Kansas
NIH	National Institutes of Health
NKVD	Soviet Secret Service (Pre-KGB)
NSF	National Science Foundation
Oranur	Orgone Anti-Nuclear Radiation
OBRL	Orgone Biophysical Research Laboratory
OEB	*Orgone Energy Bulletin*
SAPA	Sand-Packet, a type of microscopic bion vesicle
SBHM	*Sexual Behavior in the Human Male,* Kinsey
SBHF	*Sexual Behavior in the Human Female*, Kinsey
SA	Sturmabteilung - Nazi stormtroopers
SexPol	Sex-Political Organization or Movement
Venona	US Army Signal Intelligence Service, which intercepted encrypted Soviet messages
VPS	Vienna Psychoanalytic Society
WLSR	World League for Sexual Reform
WRMO	*Wilhelm Reich Mysteries of the Organism*, film
ZPPS	*Zeitschrift fürPolitical Psychology & Sex-Economy*

*"Where they burn books,
they will ultimately burn people also."*

\- Heinrich Heine, 1821
German Jewish poet

Preface and Acknowledgments

The idea and necessity for this book developed around 2011, when a public tidal wave of slander swept through the mainstream media, aimed at destroying the research legacy of the late Dr. Wilhelm Reich. The opening shots in this new war of defamation occurred with the publication of the Christopher Turner book *Adventures in the Orgasmatron*. It then became a matter of professional responsibility, to the facts and truth, to more fully investigate and publicly oppose those attacks, and to *stand up for Reich*. As a working scientist and former university professor who had evaluated and confirmed at a high level many of Reich's central theories and experimental findings, I was in a good position to take on the task.

For over 80 years, Reich has been subjected to deadly attacks by various Nazis, Communists, psychoanalysts and psychiatrists, Marxists, Stalinists and Comintern spys, and by various characters openly supporting the Kinsey agendas of hypersexual pornographic license, merged with the hard-left politic. A few Christian moralists also attacked, but hardly sufficient to justify the usual accusations (hurled from the left) that Reich was destroyed by "right-wing American McCarthyites", or the like. That consensus illusion is today nearly as entrenched as is the larger public's ignorant acceptance of the false narrative of Reich's life and work as promoted by his detractors, which are additional reasons for recording the facts into this book.

Personally I had considered, until most recently, that the worst of the mudslinging slander directed against Reich was diminishing. Reich had been dead for over 50 years. New experimental investigations of his discoveries, with solid confirmations, had been undertaken by physicians and scientists in both America and Europe. These investigations proved Reich's clinical sex-economic and orgone-biophysical discoveries held great merit. A reconsideration of Reich was quietly underway in universities, by younger generations of scholars. However, the most recent attacks shattered the calm with a renewed storm of lies and clever

In Defense of Wilhelm Reich

deception, appearing in multiple "top" publications. This demanded more than the usual rebuttal letters and articles, if only to set the record straight for history, and for the children of the future.

In addition to examining and refuting the specific charges within multiple slander articles directed against Reich, I have assembled many lesser-known facts about his detractors from published or archive sources, often with elaborated details in the References section. This includes new documentation from Soviet archives, of Reich's name appearing on a c.1936 NKVD death-list, and about an extreme left-wing "civil liberties" Judge sitting on the US Supreme Court, whose legal opinions were cited as justification for ruling against Reich on appeals, and for burning his books.

For this undertaking I consulted a great many published books and articles, as well as unpublished materials from various archives. These centrally included Reich's original books and research journals, which had already been in my personal library since c.1970. I consulted materials from both his European and American periods of work, his *Conspiracy* volumes and court trial records, including all those materials once ordered literally "banned and burned" through a court-approved *Food and Drug Administration* (FDA) injunction. Over the years I also made several trips to the *Wilhelm Reich Museum* in Rangeley Maine, and with the kind permission of Mary Boyd Higgins and Kevin Hinchey, Directors of the *Wilhelm Reich Trust*, I reviewed some of Reich's personal and unpublished documents at the *Reich Archives,* currently located at *Harvard Countway Library*. My investigations also included a Freedom of Information Act search of both the FDA's and the FBI's files on Reich, including declassified FBI files on Soviet spy rings, once marked "Top Secret". I also reviewed materials from the J.B. Matthews archive at *Duke University* on the Consumers' Union, as well as published documents from the mid-20th Century Senate and House hearings on un-American activities, and which were aimed at rooting out Nazi, KKK and Communist subversives within the American government. The names of Reich's central detractors frequently appeared in all these types of documentation.

An investigation was also made of archive materials from the *National Institutes of Health* deposited by Aurora Karrer, Reich's last wife. I also knew Eva Reich over many years, the daughter of Wilhelm, and prior to her death had conversations with her and with other physicians and natural scientists who had known and

Preface and Acknowledgements

worked with Reich, such as Bernard Grad, Morton Herskowitz and Richard Blasband, on the subjects detailed herein.

Especially, I wish to thank James Martin for his prior years of investigative research summarized in the important book *Wilhelm Reich and the Cold War* (soon to be republished). Martin began his work anticipating he would *not* confirm Reich's belief that his primary attackers were Communists. But the acquired evidence was overwhelming, that Reich's detractors were serious *Reds*, including several who were contacts or members of major Soviet spy rings. His findings clearly identified the fingerprints of hard leftists and communist operatives in the destruction of Reich, and my own investigations as presented here in large measure were an exercise in double-checking, and ultimately confirming Martin's prior historical findings.

I wish to also thank Peter Nasselstein, Tom DiFerdinando, Stefan Müschenich and Leo Köver, who provided very helpful critique, proof-reading and additional documentary materials and clarifications, notably on Reich's European period of work. Also my thanks to Gary Douglass, who gave the book a final proof-reading. I also extend my thanks to John Wilder, Bernd Laska, Richard Blasband, Courtney Baker, Myron Sharaf and Jerome Greenfield, for their additional prior research investigating Reich's biography and detractors, and related issues. Their published findings are listed in the *Additional Materials for Consultation* at the end of the Reference section. All these individuals provided numerous small and large helps in my own investigations. I also thank Jon East, writer and director of the short dramatic film about Reich, *It Can Be Done,* for his helpful advice in navigating my official complaint against a British media smear aimed at both Reich and myself, as described in Chapter 5.

Finally a heartfelt thanks to my wife Dasa Brückner. Her personal knowledge of Wilhelm Reich and 20th Century German history was of great help, and she translated many German-language documents necessary for this work to be accomplished. She also helped edit the English manuscript for this book, and provided encouraging support throughout.

James DeMeo, PhD
Greensprings, Oregon, USA
Spring 2013

In Defense of Wilhelm Reich

Wilhelm Reich, M.D. 1897-1957

1. Introduction

*"Love, Work and Knowledge are the Wellsprings of Life.
They should also govern it."*
Wilhelm Reich

Dr. Wilhelm Reich is the man whom nearly everyone loves to hate. No other figure in 20th Century science and medicine could be named who has been so badly maligned in popular media, scientific and medical circles, nor so roughly abused by power-drunk federal agencies and arrogant judges.

Publicly denounced and slandered in both Europe and America by Nazis, Communists and psychoanalysts, placed on both Hitler's and Stalin's death lists but narrowly escaping to the USA, subjected to new public slanders and attacks by American journalists and psychiatrists who deliberately lied and provoked an "investigation" by the US Food and Drug Administration (FDA), imprisoned by American courts which ignored his legal writs and pleas about prosecutorial and FDA fraud, denied appeals all the way up to the US Supreme Court, which rubber-stamped the FDA's demands for the *banning and burning of his scientific books and research journals,* and finally dying alone in prison – who was this man, Wilhelm Reich, and why today, some 50 years after his death, does he continue to stir up such emotional antipathy? It is a literal *80-Years' War* of continuing misrepresentation, slander and defamation.

Who were and are his attackers? And what stands behind their obsession to utterly rid the world of him and of his work?

Reich was a physician and natural scientist, a protege of Freud, discoverer of the *function of the orgasm* as an electrophysiological discharge and biological regulatory function. He identified how early childhood traumas and the suppression of both maternal-infant love and later adolescent peer love (i.e., the romance of Romeo and Juliet), leads to neurotic and even sadistic/masochistic behavior, with adults who grew up under such hard conditions

In Defense of Wilhelm Reich

frequently collapsing into violent fascist ideologies. Reich also was a therapeutic innovator, social reformer and antifascist freedom fighter, investigator of cancer biophysics and the origins of life, discoverer of a specific *life energy*, the *orgone energy*, and inventor of an apparatus which could concentrate this same cosmic force from the atmosphere, the *orgone energy accumulator*. From this he developed new methods to trigger rains in droughty or desert regions. He wrote about the nature of cosmic space and gravitation, unifying biology and atmospheric science with astronomy in a comprehensive theory on *life energy functions in nature: Orgonomy.*

In subsequent chapters I will summarize Reich's discoveries and provide details on those who mercilessly assaulted him for his findings. A significant library of books has already been written by Reich and his various associates and followers, and I will acquaint the reader with the best of them. Certainly it was and remains controversial. So it is not surprising that he would "kick up dust" within carefully guarded academic and medical circles – although that is a rather mild description of the reactions against Reich's work and ideas. Covering this controversy will be the major task of this book, *In Defense of Wilhelm Reich*, though I am forced to summarize too much into too-few pages, and so do not pretend this will be a thorough discussion on the exacting details of Reich's work. The primary goal will be, to educate and illuminate about the basics of Reich's discoveries and how his detractors reacted to them. I will expose their reactionary ideologies and irrationalism, their errors in understanding, as well as their deliberate lies and slanders as circulated within their public writings. Also the betrayals by several of his close associates will be discussed.

There were the early 1930s attacks by the European psychoanalysts, who objected to his therapeutic innovations and social activism, and through the usual back-stabbing and scheming, threw him out of their organization. Then the attacks from the Nazis, for his daring to expose their underlying sexual psychopathology and the irrational fanaticism of their followers. Then also, the German Communists attacked. He had actually worked with the Marxist parties towards reforms of family law, and in opposition to growing Hitlerism. However, they also became offended at his writings which highly valued basic sexual freedoms – such as the right to premarital love, to pick your own mate, to have access to contraception, or to divorce, at a time when most were illegal. The Marxists complained Reich was "distracting youth from the class struggle", and objected to his critiques of Marxist theory, which increasingly robbed everyone of both property and freedoms. In the end, they also threw him out of their organization.

The Communists and Nazis banned and/or burned Reich's publications, and both issued orders for his arrest, placing his name on their respective death-lists. The European psychoanalysts and psychiatrists, meanwhile, forbade his writings in the journals they controlled, which at one time had graced nearly every issue they produced. They also slandered Reich in "whispering campaigns" of malicious gossip, aiming to destroy him professionally. This was done, even as they made deadly compromises with the Nazis – demanding Jewish members to resign, for example, so their German organization could better "get along" with the Nazi regime.[1] In the end, many psychoanalysts also had to flee, or risk arrest and death in concentration camps.

Throughout all this, Reich kept *working, working*, making new observations and findings, publishing new books and articles, his ideas attracting a significant following wherever he landed – which upset his critics all the more.

In early 1933, Reich fled to Scandinavia where he was graciously helped by Danish and Norwegian analysts and scientists, who gave him shelter. He was also granted laboratory space at the University of Oslo, where he undertook what became breakthrough bioelectrical experiments on human subjects, the very first objective measures of human emotional and sexual excitation, which unexpectedly led to new microbiological work clarifying the *origins of life* question. These discoveries garnered him much appreciation within small circles of honest scientific investigators, but various Danish and Norwegian newspapers controlled by the Nazis and Communists launched smear campaigns against him, endangering his life. Psychoanalysts were also happy to join in with the slanders. Reich fled west in 1939, to America, as did many of his Freudo-Marxist slanderers and attackers, all on the run from the Nazi death machine, then allied with the Marxist death-machine, controlled by Comrade Stalin.

Once in America, Reich independently reconstructed a new laboratory and clinic, and secured his existence as physician, natural scientist and lecturer. His focus was then fully aimed at investigation of the new life-energy phenomenon he had previously

In Defense of Wilhelm Reich

identified in Norway. But the controversy continued, especially after his newer findings were published. In the 1940s and 1950s, new whispering campaigns irrupted from the psychoanalysts and psychiatrists. Additional lies and smears were published in American magazines, newspapers and books. These were written by American Stalinists, who frequently had longtime associations with his European detractors. But Reich continued *working, working*, and publishing new English-language editions of his older works and newer experimental findings.

Reich had identified the bioelectric currents in the body as an expression of a deeper-lying and powerful *life-energy*, which was brought into the organism through breathing, and in foods and water, and by direct skin absorption. He later called it the *orgone energy*, a phenomenon that existed freely in the open atmosphere, and was attracted to matter, either weakly or strongly, and accumulated within matter at different intensities. He developed a special chamber to accumulate this atmospheric-biological energy to a higher charge within the interior, composed of alternating layers of ferromagnetic metal and dielectric materials, much like a *Faraday cage* or *hollow capacitor*. Organisms charged up inside the accumulator showed an increase in vitality. Plants increased in growth when charged inside. Wounds would heal more quickly. Skin tone and perception were enhanced. Low-energy syndromes and illnesses reduced in intensity as biological energy level was increased. Cancer mice lived longer when charged inside the orgone accumulator, by up to three times the control group of uncharged mice. Human cancer patients reported beneficial effects. New groups of physicians and other professionals were attracted to these findings. They came to study and train with Reich, to learn the new science of *Orgonomy* which he had developed, and they all published their findings in specialized books and journals.

These new activities unhinged his American critics, who boiled with rage all the more. Unfortunately, Reich's critics were "well-connected", even "popular", and knew "all the right people" in high society, media and government, while Reich as a serious scientist avoiding the limelight was not, and did not.

New slanders aiming to destroy Reich then appeared in the popular press, and spread like wildfire in what Reich later termed an *emotional chain reaction*. He was attacked by dirty-minded writers with sexual slander, for his clarifying work on the *Function*

4

of the Orgasm, *The Sexual Revolution,* and *Mass Psychology of Fascism*. These works not only criticized the church moralists who had for millennia kept women in a low or even slave status, with their opposition to contraception, abortion and divorce, but Reich's works also criticized the Stalinists and Marxists for their betrayals of human freedom in the Soviet Union and elsewhere. Predominantly Marxist-Stalinist writers cloaked as "consumer activists" wrote lying slanders of Reich's therapy methods and the orgone accumulator as something sinister and salacious. They falsely accused him of "masturbating patients" and using the orgone accumulator as "a sex box".

The worst of the slanders were reprinted in unethical medical journals by some of the same compulsively-perpetually angry psychoanalysts and psychiatrists, who also were frequently cloaked Marxist-Stalinists. "Reich is a quack", "Reich is a danger to patients", "Reich should be investigated", they whispered or shouted. Reich and his associates did their best to counter the smears with their own published articles and legal approaches – his associate Dr. Theodore Wolfe wrote a small book, *Emotional Plague Versus Orgone Biophysics*[2] addressing the worst of it – but nothing worked to halt the onslaught of slander and defamation.

Reich's well-connected critics put the slander articles into the hands of "consumer activist" government agencies. He was then subjected to a malevolent FDA "investigation", which used the yellow press slanders as their source for "facts", and aimed to *Get Reich* on whatever pretense they could. Using the slander articles as a starting point, they began by looking for a "sex racket" and the use of the orgone accumulator as a "sex box", but could not find any evidence along those lines. Then they switched tactics, and relied upon equally cooked-up lies, that Reich was supposedly selling the orgone accumulator as a "cancer cure". No evidence existed to support that either, but this did not deter the FDA. They took a few of Reich's statements out-of-context, and obtained negative opinions about his work from a few FDA-compliant scientific and medical consultants, who either tested the orgone accumulator for miracle cures, or made mere armchair analyses of Reich's writings.[3]

The FDA and a compromised prosecutor who once had been Reich's personal attorney, then together perpetrated a fraud within the US courts, by introducing biased-bogus FDA "evaluations" of the accumulator, and anti-Reich slander from the popular press as

In Defense of Wilhelm Reich

"fact". They asserted, the orgone energy "does not exist", and therefore the orgone accumulator was "mislabeled merchandise" falling under the domain of the FDA's regulatory power over drugs and cosmetics. They also declared, Reich's published books and research journal articles were "advertising literature" for sale of the orgone accumulator. While this fraud perpetrated by the FDA and agreed to by Reich's unethical former attorney was bad enough, all the judges who ever heard or reviewed Reich's case also played their role by agreeing to the fraud, and in approving the *burning* of Reich's writings.

The first judge on his case openly welcomed the FDA's fraudulent *Complaint for Injunction,* but *refused to accept Reich's submitted written Response* which exposed the fraud and lies, basically discarding it because it wasn't presented to the court in the most formal and "acceptable" manner. Having discarded Reich's *Response,* the judge then granted the FDA's Injunction, giving them everything they wanted towards destroying Reich and erasing his existence. The judge ordered Reich's books and research journals to be *banned and burned,* in a judicial decree sounding like it came from a totalitarian dictator. It was a staggering attack upon the First Amendment of the US Constitution. Six tons of Reich books and research journals were later hauled off by FDA agents for burning in New York City incinerators, in what would become *the most outrageous incident of government-ordered book-burning in American history*.

When a technical violation of the FDA Injunction occurred in the moving of some of the "forbidden" books and research instruments across a state line by one of Reich's assistants (without Reich's knowledge), they were both arrested for "Contempt of Court". During the ensuing Contempt proceedings under a second judge, Reich was dragged into court in handcuffs and *forbidden* to reintroduce his original *Response,* or otherwise to speak about scientific evidence in defense against the original FDA and prosecutorial lying and fraud. The new judge basically muzzled Reich, and expressed no concerns whatsoever about the issue of book-burning and prohibiting scientific books from crossing a state line. Open discussion was forbidden regarding anything which was relevant to how such an alarming breach of American Constitutional protections could have happened in the first place. The push to *Get Reich* thereby continued. Under such a cloud of arrogant judicial malpractice, Reich was predictably convicted of "contempt" and sentenced to two years in prison. He then took his case on appeal, all the way to the US Supreme Court.

Throughout this gut-wrenching legal meat-grinder, Reich continued *working, working*, making new experimental discoveries and writing ever more articles and books, with new important scientific breakthroughs. His findings on the biological and atmospheric orgone energy were expanded into a broader understanding of cosmic energy, existing in the open reaches of space. He documented the existence of orgone energy in high vacuum tubes. He charged up Geiger counters inside strong, room-sized orgone accumulators constructed at his new laboratory in the forests of rural Maine, and got them to sing with very high counts per minute using only background radiation as a source. He discovered the mutual attraction between orgone energy and water, and their relationship to electrostatic phenomena. He developed a larger theory of *Cosmic Superimposition* whereby the spiral forms observed in the motions and structures of microscopic organisms was related to those of seashells and in tree branches, related to the sexual joining of male and female in the creative process, and finally as seen in the formation of hurricanes and galaxies. He developed a device, the *cloudbuster*, whereby cloud growth and rainfall could be influenced, to end droughts and green deserts. He made serious steps towards detoxification of nuclear materials. He got a small motor to turn, powered only by the background medium of the vacuum orgone energy.

None of his new discoveries mattered to those writing public slander against Reich, as they distorted and abused everything to declare him "insane", or a "dangerous crackpot." His Marxist-Stalinist slanderers were then joined by a few conservative Christian slanderers, in addition to the old psychoanalytic and psychiatric slanderers, and new pop-media slanderers. They all continued to spread their poison, laced additionally with schadenfreude and gloat over Reich's legal predicament.

Regional appellate courts, and eventually the US Supreme Court, arrogantly refused to review Reich's case and also ignored the issue of book-burning, thereby affirming the legitimacy of burning scientific/medical books and research journals in the USA, and of imprisoning a scientist and physician for trying, however imperfectly, to defend his research findings. His appeals exhausted,

In Defense of Wilhelm Reich

Reich was incarcerated in Lewisburg Federal Penitentiary. He died less than a year later, in November 1957, shortly before his anticipated parole hearing.

Reich had been publicly slandered and railroaded to his death, through professional negligence or deliberate malpractice as committed by various physicians, scientists, journalists, bureaucrats, prosecutors and judges. Especially the various judges, by their actions, expressed a severe contempt for the US Constitution's guarantees and protections, in an arrogant display of raw judicial power over vital legal and moral principles – including their sworn oaths to "protect and defend the Constitution". On their watch, they allowed to happen in America what had previously occurred only at the hands of the Nazis and Communists – except for the fact that Reich had escaped the European Nazi and Stalinist death-squads. He chose to stand and fight in America, which betrayed him and its own better nature and moral principles.

Reich's attackers were not satisfied, however. The same intensity of hatred and malicious public lying and slander about his work has continued over the decades since Reich's death, carried forward on a blizzard of copycat slander articles which invented additional lies. His new critics appear decidedly happy about the book-burning, which they pass over merely as an irrelevant footnote, specifically as one finds within the "skeptic clubs" and their dominant mainstream media supporters.

While Reich's supporters came from all different walks of life and political backgrounds, in fact *the overwhelming majority of his public detractors, both during his life and in the years afterwards, have been either leftists or outright Marxist-Communists of one or another variety.* Several were members of Soviet spy-rings or front groups. Only a few have been Christian conservatives or "McCarthyites", unlike what is frequently claimed. One primary work detailing the facts about Reich's detractors is James Martin's book *Wilhelm Reich and the Cold War,*[4] which gives abundant documentation. I've given a summary of similar material in a prior article *"New Evidence on the Persecution and Death of Wilhelm Reich"*,[5] but will provide greater detail within this work. I will also identify yet another emergent group of Reich-detractors, notably open or cloaked propagandists for pornography and pedophilia, working within the decidedly Stalinist "skeptics" groups and literary salons. They appear motivated by a severe hatred of Reich's work favoring *heterosexual genitality*[6] and the *function of the orgasm*, as well as for his later anticommunism.

Nearly all of Reich's detractors were deeply bothered by his findings on sexuality, and with his *SexPol* social-political work that aimed to legalize contraception, abortion and divorce, and give women equal rights to men. His assertion that childhood masturbation was harmless or even beneficial to later sexual development, his endorsement of responsible premarital sexual love among peers, as with the young Romeo and Juliet, elicited great outrage among moralists within the Church, and among both the Nazis and Communists. He also spoke against pornography and prostitution as the consequences of sex-repressive society. He further viewed homosexuality and sadomasochism to be curable neuroses, and demanded the protection of children against adult seducers and pedophiles. History shows, Reich's emphasis upon heterosexual genitality would not be accepted either by those pushing Church-demanded premarital sexual abstinence, nor by the anti-Church factions demanding that "polymorphous perversity", homosexuality, bisexuality, sadomasochism and so on, should be accepted as equal to heterosexuality. Especially the *Kinsey agenda* of sexual promiscuity and an equivalence of all different kinds of sexual expression would ultimately find followers who loathed and attacked Reich's findings. With Reich's death in prison, the burning of his books, and the popularized but *completely false "discrediting" of his clinical and scientific work*, the Kinsey supporters largely won a public debate in which Reich's ideas were never allowed into discussion.

Additional details on these socially burning issues are given in the pages to follow, which primarily expose how Reich is today slandered with lies as being some kind of "sex-monster", generally by people who openly or secretly promote their own sexual monstrosities. Clearly, Reich's findings on genitality, his conversion from young European Marxist into American anticommunist, and his ideas on cosmic energy have upset a lot of people who just can't leave the man alone.

Beyond the absolute number of the *Get Reich* attack-and-smear articles,[7] and their continuation now for over *eight decades*, it is the sheer *intensity* of the maliciousness directed against him, the bitter-mouthed lying towards the goal to utterly destroy the good name and entire life-work of a man long dead, the intensely personal

In Defense of Wilhelm Reich

nature of the attacks employing the most outrageous lies, omissions and half-truths, which stuns even those who know little about him. It is a *protest too much* by a conspiracy of dunces, *angry little men*, to use Reich's terms. Even today, with all the self-proclaimed "intelligent sophistication" and "openness to new ideas and scientific findings" – which is often a cover for back-stabbing censorship and repression of the truly new discoveries – Reich continues to *deeply, personally and bitterly offend*.

The late Dr. R.D. Laing caught the essence of it in 1968:[8]

> *"He assaults our narcissism in almost unforgivable ways. Freud was cool. Reich is uncool. He tells us that homo normalis is a sort of bladder, often dried up, sometimes overtaken with convulsions, longing and terrified to burst, whether through penetration from outside, or explosion from within: terrified to live freely, which would be to love; with an insane fear of being destroyed and at the same time with a senseless readiness to die, to destroy what he fears; fearful of almost everything, most of all, himself; psychically turned inside out, persecuting righteously his persecutors outside himself, none other than his own projections of evil."* [8]

I was just a few years out of high school and only beginning to read Reich when Laing wrote those lines. It was a liberating experience to finally read someone who understood the sexual problems of youth, who spoke to the heart, emphasized love over pornography even while accepting the intense sexual pressure felt most urgently by the young, a man who offered an understanding of the neurotic and frequently violent nature of many families, teachers, doctors, priests, government officials, and so on.

Reich did more than merely identify the problems, however. He provided *a pathway out of the deadly social traps,* and that made all the difference. So I was never totally surprised at how the "uncool Wilhelm Reich", and anyone who dared to take him seriously, was always getting mercilessly hammered by their "open-minded liberal understanding" or "forgiving Christian" peers. Out of curiosity, years later I assembled a citation list of over 100 published articles and books which attacked Reich, mostly from the "top" newspapers and national magazines, and it is hardly a complete listing.[7] *And it still continues today!*

Introduction

In Defense of Wilhelm Reich will firstly present an abbreviated time-line overview of his biography and major research findings and publications, which includes many details on historical context and about his detractors. This is followed by an equally abbreviated overview of the more significant scientific studies which have confirmed Reich's controversial findings. Then will follow a number of specific rebuttal articles standing against the most significant and recent public slanders of Reich. Especially the critique given for the Turner "Orgasmatron" book will go into some depth of discussion on the specific key points of Reich's clinical findings on sexuality.

Objectively-determined facts and truth should guide people's decisions and world-views, "popularity" or "political correctness" be damned. Mostly, they do not. The self-blinding nature of human armoring and rigid character structures has led our species to run from its own biological core and emotions, to praise the rationalizing intellect at the expense of the emotional heart and natural physical longings for love and passion, to squash down core feelings and uplift every kind of emotionally plagued individual to act as petty bosses or Grand Dictators over our lives. This has badly retarded the causes of human freedom and taking of personal responsibilities. It has diverted scientific inquiry down blind alleys, leading to a rise of Medieval thinking within the universities and hospitals, with alarming official approvals for censorship, including the public character assassination and imprisonment of heretics, and the burning of their books.

In Defense of Wilhelm Reich

2. Wilhelm Reich Timeline: Biographical Summary, Discoveries and Attacks *

For full citations to the papers and books mentioned in this section, see the online *Bibliography on Orgonomy*
http://www.orgonelab.org/bibliog.htm
For full citations to the materials attacking Reich, see:
http://www.orgonelab.org/bibliogPLAGUE.htm

Reich's European Period

1897: Wilhelm Reich is born on March 24 in Galica, a province of the Austro-Hungarian Empire. He is home-birthed, and home schooled. His childhood is spent on the family farm in Bukovina, with a private tutor for sciences. This same year, Freud is revising his early work, rejecting his own early observations about real childhood sexual traumas (rape, incest) as the cause of neurosis and other emotional disorders, replacing it with a new "seduction-wish theory".

1910: Reich's mother commits suicide after her husband learns of her affair with one of Reich's tutors.

1914: Reich's father dies of pneumonia, leaving him in charge of the family farm at age 17. World War I begins.

1915: War-chaos. Russian troops invade the Bukovina region after Austrian troops invade Serbia. Reich's family estate is abandoned and destroyed. Reich flees west with his younger brother Robert. He joins the *Austro-Hungarian Army* as an artillery officer. Between 1915-1918, he is active on the Italian battle front.

---

* Expanded from the outline of a 16-week university lecture series developed and presented by the author since c.1975. A more detailed biography of Reich is found in Myron Sharaf's *Fury on Earth*.[1]

In Defense of Wilhelm Reich

1917: The February 1917 Russian Revolution leads to the abdication of the Romanov Tsar and establishment of the *Russian Provisional Government* by the former *Imperial Duma*, in concert with the new *Democratic Soviets*, or village-level councils. In April, the German Kaiser and his military High Command sponsor Lenin's secret train ride from Switzerland to Petrograd, and finance his push to destabilize Russia and take it out of the war. In October, Lenin's Bolsheviks shoot their way into power, betraying the February Russian Revolution and originally-democratic Soviet village and neighborhood councils. Mass executions of political opposition and "counter-revolutionary" civilians begins, led by the Cheka and the Red Army with Trotsky in command. Massacres and cover-ups abound. The Soviet border is thereafter sealed and newspapers are censored, with massive propaganda to depict a "worker's paradise".

1918: WW-I ends in November. The *Treaty of Versailles* demands Germany and Austro-Hungary dismantle their military down to nominal levels. Reich leaves the army to settle in Vienna as one of many impoverished war-refugees. He begins to study at the *University of Vienna, Faculty of Law,* but finds this unsatisfying, and soon transfers to the *School of Medicine.* The social atmosphere in Vienna is lively and intellectually rich. Reich excels in his medical studies, and soon begins to tutor other students.

Lenin, now firmly in power, orders the murder of the incarcerated Romanov family.

1919: Reich joins an informal student seminar devoted to sexology and other new ideas. He is soon elected leader of the group, and is reading the early works of Sigmund Freud, such as *Sexuality in the Aetiology of the Neuroses* (1898), *Three Essays on the Theory of Sexuality* (1905), and *Civilized Sexual Morality & Modern Nervousness* (1908). Reich meets with Freud and the contact between them increases. He also has classes with biologists of the vitalist school, Otto Warburg and Paul Kammerer.

Sigmund Freud

1920: Reich is admitted as a guest member in the *Vienna Psychoanalytic Society* (VPS). He presents a paper to the VPS on Ibsen's *Peer Gynt,* and publishes the semi-autobiographical *"Breakthrough of the Incest Taboo"* [2] in the psychoanalytic press. Shortly afterward, Reich is elected as a full member. Freud, meanwhile, is retreating from his earlier positions, and publishes *Beyond the Pleasure Principle,* arguing for a specific "death instinct". Reich never accepts this idea.

1921: Reich continues to study for his medical degree, and presents many new papers to the VPS. With Freud's approval he begins treating patients psychoanalytically.

Bolshevik power in Russia is unsuccessfully challenged by the Kronstadt sailors, given Lenin's betrayal of promised democracy. *"Soviets Without Communists"* is the sailor's slogan. Trotsky's Red Army attacks and overwhelms the sailors, massacring hundreds who surrender, as well as many of their civilian supporters and families, in several days of drunken bloodlust and looting. (Trotsky later denies responsibility.) Information on these events is suppressed, and published only years later in America or Britain by refugee Russian survivors, or decades later by historians following the collapse of the Soviet Union.

Hitler becomes the leader of the *National Socialist German Worker's Party* (Nazis).

1922: Reich receives his medical degree from the University of Vienna, and begins postgraduate study with Wagner-Juaregg, and Schilder. He marries Annie Pink, who also becomes a psychoanalyst, and begins 8 years of work with the *Vienna Psychoanalytic Polyclinic,* where he eventually becomes Assistant Chief. This period marks the start of Reich's work on the primacy of *genitality,* development of the *orgasm theory,* new techniques of *character analysis,* and critique of the *impulsive character.*

Lenin and German militarists forge secret agreements to build new large munitions factories deep in Soviet territory, to evade and defeat the Versailles disarmament agreements.[3] These factories employ German engineers to build new tanks, artillery, fighter aircraft and bombers, along with chemical weapons. German troops are also secretly trained in their joint planning for new wars of conquest. The Soviet Red Army led by General Trotsky is also on

---

Reich's New Ideas in Psychoanalysis, starting c.1923.

- Neurotic symptoms are only the "tip of an iceberg" which extends subconsciously to involve the entire character structure, in a complex of psychological defenses and attitudes;

- A therapeutic focus upon *analysis of resistances*, rather than only of information content revealed in therapy;

- Analysis of the *form* and manner of behavior, the role of body language, muscle tension and respiratory blocking;

- The concepts of *character armor and somatic-body armor* as defenses to bind up anxiety, forbidden feelings and painful memories associated with *real past traumas*;

- Elucidation of *natural primary versus diverted or bottled-up secondary drives, genital versus pregenital impulses,* and *self-regulated versus compulsive behavior and morality*, with a distinction between the organismic *core versus periphery*.

- A delineation of character types and a revision of analytical methods and approaches;

- Criticism of patriarchal authoritarian family law which reduces women and children to chattel; a critique of sexual taboos and phobias which lead directly to neurosis and mental illness.

---

the march, brutally suppressing dissent within Soviet-Russian territory. Lenin orders Trotsky to invade Poland, as part of their violent international "war communism". No longer fighting unarmed peasants, the Red divisions are routed by the Polish Army.

1923: By this date the 26-year old Reich has presented many papers to the VPS and published several of them in psychoanalytic journals, mostly on the subject of childhood sexuality and adult sexual phobias and neuroses, with many reviews of publications by other analysts. He now publishes *"About Genitality"*.[4] Freud at age 50, meanwhile, is further retreating from society, contracts cancer of the jaw, and harbors mixed feelings about the young Reich.

Hitler and his Nazi allies stage a failed coup d'etat in Munich. Hitler is imprisoned for 8 months.

1924: With Freud's approval, Reich organizes a *Technical Seminar* for young students in psychoanalysis. He publishes *"The Therapeutic Importance of Genital Libido"*,[5] and introduces the concept of *orgastic* *potency.* By the end of this year, Reich has published no less than 38 separate articles and book-reviews on the subjects of human sexuality and therapeutic methods,[6] emphasizing his own ideas on genitality. These are mostly published in the major psychoanalytic journals such as *Int. Zeitschrift für Psychoanalyse* (IZP), *Imago,* and *Zeitschrift für Psychoanalytische Paedagogik.* He is now a leader in the psychoanalytic movement and organization. Reich also joins the *Austrian Social Democrat Workers Party* (APÖ). Eva Reich is born to Annie and Wilhelm Reich.

Lenin dies, victim of an earlier assassination attempt and stroke. A struggle ensues between Trotsky, who pushes for global *"war communism"*, and Stalin, who works for consolidation of communist dictatorship within the USSR as a first step.

1925: Reich publishes: *"The Role of Genitality in the Therapy of the Neuroses",*[7] and *The Impulsive Character.*[8] In these, he argues for a single systematic theory of character; that a permissive childhood with later traumatic sexual blocking leads to impulsiveness; that the impulsive disorder is more severe than neuroses, but less so than psychoses. These works maintain many traditional analytical concepts which he would later reject, such as latency in childhood. He also writes *"More Comments on the Therapeutic Importance of Genital Libido".*[9]

Hitler publishes *Mein Kampf.* Trotsky loses the power struggle with Stalin. He is eventually exiled from Russia and begins writing various false revisionist histories which erase his own terrorism and murder as leader of the Red Army.[10]

1926: Reich publishes *The Sources of Neurotic Anxiety,*[11] a paper marking his interest in the bioenergetics of emotion and sexuality. He continues publishing many other minor papers and book-reviews, and is an increasingly influential worker within the psychoanalytic movement. However, Freud rebukes Reich on several points of his new character analytic methods.

1927: Reich publishes *Die Funktion des Orgasmus*[12] (the modern English edition is retitled *Genitality in Theory and Therapy of*

In Defense of Wilhelm Reich

Neurosis). In this, he makes a preliminary definition for *orgastic potency*, arguing that male and female orgasm are functionally similar, noting also the widespread problems of sexual anesthesia and impotence. He emphasizes the differences between genital and pregenital sexual excitation and discharges, and the necessity for both erotic excitation and loving surrender to one's partner as prerequisites for full genital orgasm. He observes the absence of orgastic potency in all neurotic and psychotic patients, and defines the establishment of orgastic potency as a primary goal of therapy. He defends childhood masturbation as natural.

His paper, *"On the Technique and Interpretation of Resistance Analysis"*,[13] challenges older psychoanalytic methods. Reich argues his ideas are *"the inevitable outcome of Freud's own theory"* (on the sexual etiology of neuroses). Freud is cold to Reich's new ideas. Perhaps due to this rejection by Freud, Reich succumbs to a bout of tuberculosis, and enters the Davos Sanatorium for treatment.

After witnessing a violent police attack on striking workers in Schattendorf, Austria, where many were killed, and alarmed at growing Nazi power, Reich joins a medical group affiliated with the *Austrian Communist Party* (KPÖ) and reads Marx for the first time, in full ignorance (along with the bulk of the European Left) of the widespread massacres going on at that time within the Soviet "worker's paradise", and of Soviet collusion with the German militarists. He considers the development of a *political psychology*, merging psychoanalysis with Marxism, and begins writing *People in Trouble*,[14] on the difficult social-sexual conditions and people's general fear of authority, freedom and responsibility.

1928: Reich publishes new papers: *"A Criticism of Recent Theories of the Problem of Neurosis"*,[15] *"About Genital Self-Satisfaction in Children"*,[16] and *"On the Technique of Character Analysis"*,[17] all of which develop his own therapeutic insights. He begins what is eventually termed the *SexPol* (sex-political) movement, a public educational effort aiming for political reforms on sexual and family law, and for *the social application of psychoanalytic principles*.

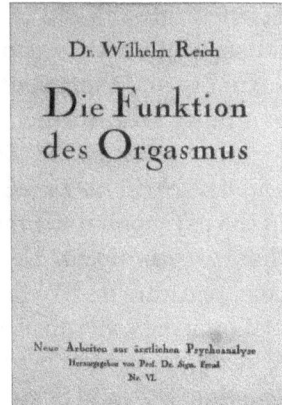

As Sharaf describes it:

> *"Energies congealed in character defenses and armor are not available for rational work or social criticism. Preoccupations with emotional problems and sexual conflicts lead to political apathy. Lack of clarity about sexual issues and, worse, lies and deceit regarding human love life undermined the capacity of people to see through political chicanery."*[18]

At year's end, Reich forms a social-outreach organization in Vienna, the *Socialist Association for Sex Hygiene and Sexological Research*. Freud advises, *"You'll be stepping into a hornet's nest"* but nevertheless supports the effort. This *Association* eventually has 4 psychoanalysts and 3 obstetricians, working with the public towards sexual counselling and education, and sexual/family reforms. A political action platform is developed, notably towards a greater freedom and economic standing for women, freedom to divorce, legal availability of contraception and abortion, advocacy for the sexual rights of unmarried youth, ending of prostitution by eliminating its causes in poverty and sex-frustration, protection of children from adult seduction, an end to distinctions between "legitimate" versus "illegitimate" children, and an ending of the Church's stranglehold on family and sexual laws.

Reich formally joins the KPÖ, hoping for a working alliance. However, Reich's SexPol ideas stand in basic opposition to authoritarian Marxist principles, in how he emphasizes sexual-emotional considerations of individual happiness and freedom, ignoring obedience to the Party and class warfare ideology.

Annie Reich, now a medical doctor, takes her training psychoanalysis with Anna Freud, a celibate (or possible lesbian) who becomes a bitter enemy of Wilhelm Reich[§] over his emphasis upon genitality as a prerequisite for emotional and sexual health.

Lore Reich is born to Annie and Wilhelm.

1929: Reich's newest articles and books include: *"Psychoanalysis in the Soviet Union"*,[19] *Dialectical Materialism and Psychoanalysis,*[20] *"The Genital Character and the Neurotic Character"*,[21] and *Sexual Excitation and Sexual Satisfaction.*[22]

Reich makes open criticism of the failures of the Austrian Social

---

§ Whenever "Reich" is mentioned alone, it always refers to *Wilhelm Reich*.

In Defense of Wilhelm Reich

Democrats (APÖ) and Communists (KPÖ) to come to grips with the deadly growth of Nazi power. (Reich is still unaware of secret and growing Nazi-Soviet cooperations.) He forms an opposition group and briefly finances a public newspaper, to sound an alarm.

In August-September Reich visits the Soviet Union with his wife Annie, but even with a limited view of what is going on, he returns with unsettled feelings. The Soviet NKVD (forerunner of the KGB) has established spy networks in Germany and Austria, to watch for dissent within the Communist Party ranks. Reich is being monitored by NKVD agent Arnold Deutsch, who works with Reich in the SexPol clinics and helps to publish several of his books. Deutsch later becomes a top Soviet spy in Britain, recruiting such as the notorious Kim Philby for the Cambridge spy ring, one of whose American members, Michael Straight, later figures centrally in the public attacks against Reich in the American press. Stalin continues with massacres, imprisonments and deportations of oppositionists into the growing network of gulag labor-death camps.

The Wall Street Stock Market crashes in October.

1930: Early in the year, Reich is expelled from the APÖ for his provocative independent organizing. Reich's paper *"The Sexual Misery of the Working Masses and the Difficulties of Sexual Reform"*,[23] is presented to the *World League for Sexual Reform* (WLSR), which highlights 700 clinical cases. In this and other lectures, he argues that compulsive marriage without sexual love and gratification, lacking contraception and with no possibility of divorce, is a deadly institution. He argues that childhood masturbation is natural, that contraception and abortion should be legalized, and that, while homosexuality is a neurosis, people should not be persecuted or put into prison for it. He also writes *"Character Formation and the Phobias of Childhood"*,[24] and *Sexual Maturity, Abstinence and Marital Fidelity: A Criticism of Civil Sexual Reform*,[25] which is critical of the limited WLSR efforts.

In September, Reich moves to Berlin, continues his SexPol work, and meets with various psychoanalysts. He forms the *Berlin Technical Seminar* and also joins the *German Communist Party* (KPD) in efforts to disseminate his Sex-Political reforms, and to oppose the Nazis.

Late in the year, Reich reads Bronislaw Malinowski's book *Sexual Life of Savages*,[26] describing the Trobriand Islanders of the

Melanesian South Seas. At that time period, the Trobrianders are a generally peaceful and non-neurotic, genitally healthy society: no rape, no child molestation, no homosexuality, no child-abuse, full adolescent sexual freedom but without sexual chaos or perversions. Monogamous love-match marriages are the norm for adults, with freedom to divorce, high woman's status, matrilineal family structure, animistic cosmic views but no priesthood.

Freud's depression and immobility over the world situation worsens; he publishes *Civilization and its Discontents*.

The Nazi Party gains in German elections, where they become the 2nd largest party in the Reichstag. Nazi SA (*Sturmabteilung*) Brownshirts make violent attacks on political and media critics. German police do little to interfere.

1931: Reich publishes a favorable book-review of Malinowski, and *"About Epileptic Seizures"*.[27] This latter paper details the biophysical arched-back characteristics of epilepsy as an extreme anxiety reaction, a kind of "extra-genital orgasm" but in full opposition to the genital-orgastic forward-bending of the body, as during the sexual embrace. Reich makes numerous lectures to various left-organizations in Berlin, at a time when Nazi SA thugs are marching in the streets and committing violent acts. In alliance with the KPD, Reich helps to form the *Unity Association for Proletarian Sexual Reform and Protection of Mothers*, through which his own *SexPol platform* was publicly promulgated. *Unity* is an umbrella organization for many other sexual reform organizations, and sponsors a major conference in Düsseldorf, where Reich's 7-point sex-economic platform is proposed for wider adoption into law.[28]

Branches of this *Unity Association*, along with sexual counselling clinics, are organized in other German cities. Some 40,000 members are eventually affiliated with *Unity*, through their various independent sub-organizations. Few or none of the freedoms or reforms detailed in Reich's SexPol platform (see next page) are legal or practically existent at the time they are proposed.

1931 ATTACKS: Karl Kautsky, an advocate for eugenics and leader of the *Marxist Socialist Equality Party* in Berlin, attacks Reich in their journal *Freiheit*,[29] accusing him of sexual propaganda against the ideals of socialist equality. Increasingly bad relations develop between Reich and the Socialist and Communist leaders and

---

Reich's 7-Point Sex-Economic Platform, 1931
Proposed at the *Unity* Conference in Düsseldorf

"1. Free distribution of contraceptives to those who could not obtain them by normal channels; massive propaganda for birth control.
2. Abolition of laws against abortion; provisions for free abortions at public clinics; financial and medical safeguards for pregnant and nursing mothers.
3. Abolition of any distinction between the married and unmarried. Freedom of divorce. Elimination of prostitution through economic and sex-economic changes to eradicate its causes.
4. Elimination of venereal disease by full sexual education.
5. Avoidance of neuroses and sexual problems by a life-affirmative education. Study of principles of sexual pedagogy. Establishment of therapeutic clinics.
6. Training of doctors, teachers, social workers, and so on, in all relevant matters of sexual hygiene.
7. Treatment rather than punishment for sexual offenses. Protection of children and adolescents against adult seduction." [28]

---

organizations, and with the moralists and dogmatists within the psychoanalytic community. The Marxists fulminate that a focus upon sexual issues will distract revolutionary youth from the struggle against capitalism. Doctrinaire psychoanalysts object to nearly everything Reich does. Psychoanalyst Sandor Rado slanders Reich, telling his wife Annie that he is insane and she should leave him. Later in America, Rado continues this slander.

Nazi policies are nearly in full opposition to SexPol. They will eventually forbid abortion among women of Germanic heritage, to boost their populations, while permitting or forcing it (and sterilization) among the "Untermensch" Jews, Slavs, Roma, etc. Women of the "Germanic super-race" would remain subordinated in the Third Reich, as per the slogan: *Kinder, Küche, Kirche* (Children, Kitchen, Church).

1932: Reich continues lecturing and publishing on various subjects in the psychoanalytic press. He founds the *SexPol Verlag* to publish his writings, now blacklisted by the psycho-analysts and communists, and soon to be burned by the Nazis. He publishes *"The Masochistic Character"*,[30] a criticism of Freud's *death-instinct theory*, and a step towards his later explanation of the subjective experience of pleasure which some neurotics get from the breaking of the skin surface, as a form of pathological non-genital release of bioenergy. He also publishes *The Invasion of Compulsory Sexual Morality*,[31] discussing the origins of patriarchal authoritarian society in sex-repression and child abuse. Using Trobriand society as a reference, he argues that *sexual latency,* the *Oedipus complex* and social violence come from those trauma-inducing processes as only found within patriarchal-authoritarian societies. Reich also publishes *The Sexual Struggle of Youth*,[32] a pamphlet for young people containing basic information on sexuality, procreation and contraception, telling them their sexual feelings are natural but widely feared by most adults. He advocates for sexual freedom and responsibility among teen peer-groups and the unmarried, similar to Trobriand society. This means allowing youth to live out their romances, to obtain and use contraception and not bring new babies into the world unless they are able to care for them, and to protect youth from sexually sick adult predators. Social irrationalism and violence would melt away, he argues, if society adopted gentle treatments for children and a more relaxed attitude about healthy heterosexuality.

Franklin D. Roosevelt is elected USA President in November.

1932 ATTACKS: Reich's new publications and lectures on sex-political subjects elicit severe condemnations by psychoanalysts, Nazis and Communists alike. He is "officially" attacked in the psychoanalytic press (IZP) by the Marxist Siegfried Bernfeld in the article *"The Communist Discussion in Psychoanalysis and Reich's 'Refutation of the Death Instinct'"*.[33] Reich writes a reply to this paper[34] which appears in the same issue of the IZP, but his

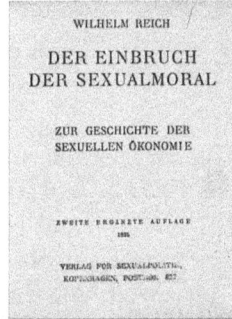

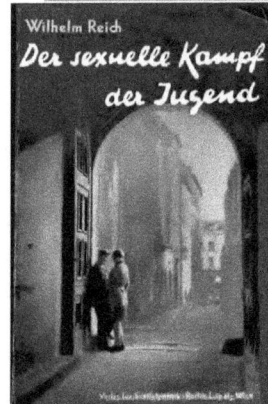

Verification and Extension of Reich's Sex-Economic Theory[35]

Reich postulated that human sadism and social violence were created by real trauma and abuse experienced by the individual in infancy and childhood, compounded dramatically by later repression of their sexuality and emotions as teenagers. He observed how the maternal-infant bond, as well as childhood sexual play and the later bonds of love between young people – exemplified by Romeo and Juliet – are always crushed out in violent patriarchal authoritarian cultures. The largest-ever cross-cultural study of human culture around the world was undertaken by the author in the early 1980s, partly as a test of Reich's theory. Human behaviors and social institutions in over 1100 different world cultures from the standard ethnographic literature were reviewed, confirming Reich's findings exactly. Societies which crush their young people with painful rituals and punitive obedience training, with strong premarital taboos, and which suppress women and sexuality, predictably have multiple violent beliefs, behaviors and social institutions. These include slavery, castes, male-favoring inheritance rules and violent "High Gods", among other similar traits. This study, detailed in the book *Saharasia*,[35] also identified the world regions of the most violent patriarchal authoritarian societies. These are found within the large desert belt laying across North Africa, the Middle East, and into Central Asia (*Sahar-Asia*), indicating the role of desertification and famine trauma in the origins of early social violence and warfare.

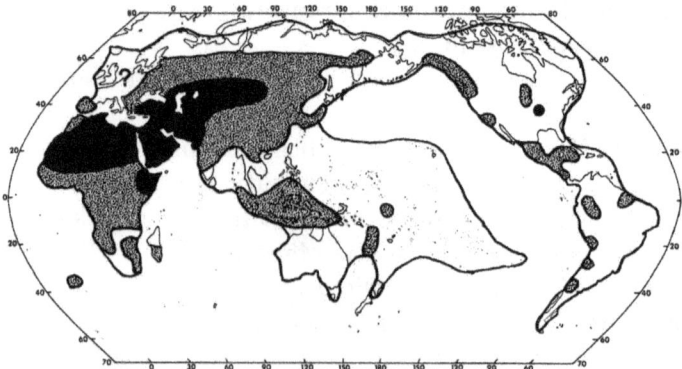

articles are no longer accepted in the psychoanalytic press.

July elections lead to widespread street violence; the Nazi Party wins 34% of the parliamentary vote.

Reich presents his *Sexual Struggle of Youth* to the KPD for distribution, but they send it on to Moscow for review and approvals, which are not forthcoming. A terse announcement thereafter appears prohibiting distribution of his books within Communist Party bookstores, in *Roter Sport (Red Sport)*, a KPD youth publication of *Fichte*, a Berlin communist sports organization:[36]

"STOP DISTRIBUTION! The pamphlets by Reich taken over by the literature distributor of the KG from the Verlag für Sexualpolitik, have been withdrawn and their further distribution is banned. The taking over of the distribution resulted from a misunderstanding. In the pamphlets by Reich, the problems are treated in a way that is contradictory to the correct, revolutionary education of children and adolescents." [36]

1932-33: Soviet-planned starvation genocide in the Ukraine kills from 9 to 11 million; Communist Parties the world over cover it up as a "drought". The *New York Times* also lies for the Soviets, publishing propaganda stories from the Stalinist Walter Duranty, denying the mass deaths – and for which Duranty wins a "Pulitzer Prize". Only smaller conservative news publications initially print the facts. Tens of thousands of "undesirable elements" are also purged from the Soviet Communist Party, some 22%, in a lead-up to the Moscow "show trials".

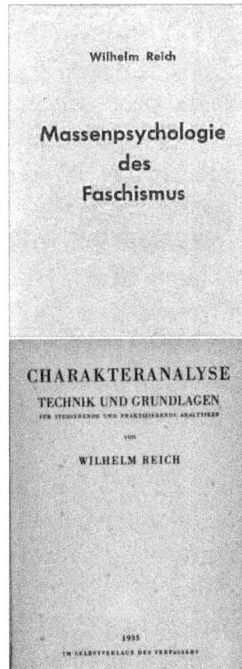

1933: With great difficulty, Reich publishes *Mass Psychology of Fascism*[37] and *Character Analysis*,[38] for which he earns great praise from some, and grave condemnation from others. He describes the Nazi movement as rooted in sex-frustration and a psychopathological helplessness and craving for authority within the average person; the authoritarian character is afraid of freedom and responsibility, and supports only those leaders who mirror their own abusive and obedience-demanding patriarchal authoritarian fathers and family structure. Sexual and political freedom creates great anxiety, and so they will vote for whomever offers them the most familiar yoke and chains. In these works, Reich also adds considerably to traditional psychoanalytic theory, marking an eventual move into his new *Sex-Economic Theory.*

Reich expands traditional free association talk-therapy and dream analysis into a more direct approach on the patient's characterological resistances. He develops new methods of *vegetotherapy*, which address the patient's *inhibited respiration* and *biophysical armoring.* This new method requires, in addition to standard character analytical methods (derived from psychoanalysis) and his own resistance-analysis, that the therapist should directly observe the patient's muscles and breathing. This requires the patient to undress down to their under-garments. Patients are then encouraged to breathe deep, to vocalize, yell and punch the analytic couch, or to kick, to feel and express bottled-up emotions such as rage, and to cry out their sadness. The new method also includes the therapist applying pressure to specific muscles where emotional tension was "stuck", such as the neck or chest muscles. This touching of the exposed muscles of the body creates a scandal among the psychoanalysts, but it gives much relief to patients. (Years later in America he calls his new approach *orgone therapy*, while decades later after his death, with greater or lesser adherence to Reich, it is frequently termed *Reichian Therapy,* or *Body-Oriented Psychotherapy.*[39])

Wilhelm Reich and Annie Reich separate and divorce, as tensions between himself and the psychoanalysts intensify. Annie Reich remains loyal to Freud and the IPA, which takes a decidedly appeasing posture towards Nazi power. Reich meets his second wife Elsa Lindenberg, a dancer who has recently lost her Berlin job due to her anti-Nazi views.

1933 ATTACKS: 29 January: KPD factions now controlling the sexual reform *Unity Association*, which Reich helped to create, openly attack him, in spite of his anti-Nazi activism:

> *"Equally monstrous is Reich's view that sexual repression occurs in all classes. In this way he denies the existence of the*

antithesis of the classes. Worst of all, however, is the fact that in 'The Sexual Struggle of Youth' he asserts that there are conflicts between generations. That means that the class struggle is to be shifted to the family instead of concentrating all forces on the political struggle against exploitation and impoverishment." [40]

January 30th: German President Hindenburg appoints Hitler as Chancellor, as a means to "restore order". Göring is appointed to lead Prussia, the largest German state.

February 27: Nazis set an arson-fire in the *Reichstag* (German Parliament) which is then blamed on the Communists. The next day, Hitler gains dictatorial powers with a new *Law for the Protection of People and State ("Reichstag Fire Decree")*. Civil liberties are suspended. SA Brownshirts are integrated into civil police, to "maintain order" for forthcoming elections.

March 2nd: Three days before the German general elections and three days after the Reichstag fire, Reich is publicly denounced in the Nazi-run Berlin newspaper *Völkischer Beobachter*, in an article titled *"Bolshevism or Germany?"*[41] Reich fears for his life and quickly flees to Vienna.

Bolschewismus oder Deutschland?

Auflösung der Familie

Zerstörung der Sittengesetze

Hunger und Tod

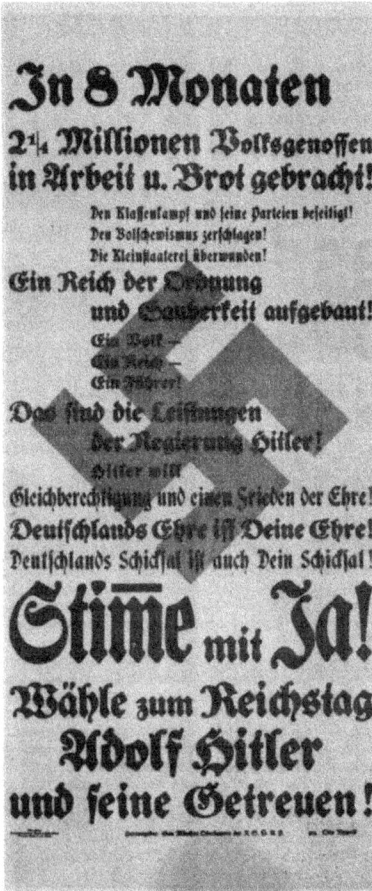

Above and Left: Nazi Propaganda Posters from the critical German election of March 1933.

Below: Nazi SA Brownshirt thugs arrest thousands of political opponents and German liberals, to no good end.

A Prussian/Nazi Expulsion Order, 5 May 1933
Prelude to arrest, detention and probable death for all listed.
Wilhelm Reich appears at #12 on the list (at bottom).

Geheimes Staatspolizeiamt.   Berlin, den 5 Mai 193 +3
    I²ᶜ 7083/81.

    1.) Vermerk: Die Bundespolizeidirektion in Wien ersucht
                    um Übersendung eines Verzeichnisses der wegen
                    kommunistischer Umtriebe aus Deutschland ausge-
                    wiesenen österreichischen Staatsangehörigen.
                    Sie bittet auch fernerhin über die Ausweisung von
                    österreichischen Staatsangehörigen um kurze Mit-
                    teilung.

    2.) Schreiben:

             An
             die Bundespolizeidirektion
                    in Wien.
             Ihr Schreiben vom 17. 4. 33 - Pr.Zl.IV-2642/33-.

                    Anbei übersende ich ein Verzeichnis der
             bisher aus dem Freistaat Preussen ausgewiesenen
             österreichischen Staatsangehörigen. Die Ausweisung
             ist wegen ihrer Betätigung in der kommunistischen
             Bewegung erfolgt. Bei weiteren Ausweisungen er-
             folgt Nachricht. Die in der Liste aufgeführten
             Personen sind bereits von selbst abgereist und
             ist der Ort ihres Reiseziels/nicht bekannt.

                    Liste
             der aus dem Freistaat Preussen ausgewiesenen
                  österreichischen Staatsangehörigen.

    1) Lilli S z a n t o , 1. 7. 10 Budapest geboren,
    2) Heinrich R o s n e r , nationale nicht bekannt,
    3) Irma S c h r ö t t e r , 26. 9. 91 Wien geboren,
    4) Karl S u r m a , 13. 2. 02 Wilmersdorf geboren,
    5) Tony L ö w e n t h a l , 6. 11. 97 Tarnowitz geboren,
    6) Alexander S t r a s s e r , 16. 11. 98 Wien geboren,
    7) Dr. Leopold K a t z , 15. 11. 89 Upsala geb.,
    8) Otto B i t t n e r , 26. 3. 03 Wien geboren,
    9) Peter K a t z , 17. 8. 04 Berlin geboren,
    10) Franz W e i m a y r , 25. 5. 07 München geboren,
    11) Josef S c h l e s i n g e r , 24. 3. 02 Günserndorf geboren,
    12) Dr. Wilhelm R e i c h , 24. 7. 97 Dobzanica geboren,
    13) Karl K o t e s n i e z k y , 27. 5. 04 Wien geboren.

In Defense of Wilhelm Reich

March 5th: The Nazi Party wins 44% of the vote, and by coalition scheming gains control of the Reichstag, leading into a full Hitler dictatorship.

March 23rd: The Reichstag is now under Nazi control and passes the *Enabling Act,* allowing the government to issue laws without further approvals from the Reichstag. Nazi SA Brownshirts begin arresting thousands of opposition political leaders and intellectuals. Max Hodann, one of Reich's physician associates in SexPol, is arrested and spends 6 months in Dachau, a recently created concentration camp for detention of political prisoners. Upon release his German citizenship and university medical degree are revoked, and he is sent into exile.[42] He is luckier than many others. The KPD and other opposition groups are abolished and go underground. Stalin remains silent on the attacks against German communists, in secret cooperation with the Nazis towards their military rearmament via Soviet weapons factories,[3] and the forthcoming *Hitler-Stalin Pact.*

April: Reich finds Vienna is also slipping under Nazi influence, and flees to Denmark in late April, where he is joined by Elsa. Reich's name appears on a Prussian-Nazi expulsion list at this time. He is a hunted man. Books by Jewish authors such as Freud, Einstein, Brecht and Reich, and many others are soon burned by the *Nazi Youth League* and SA, notably in a frightening hysterical celebration on May 10th at Opernplatz in Berlin.[43] Jewish professors, government workers and professionals are soon fired from their jobs. Annie Reich and her children flee to Prague, joining a circle led by the Marxist psychoanalyst Otto Fenichel. Annie marries Arnold Rubinstein, a one-time high level Soviet spy who hates Reich, known within the Comintern as "Comrade Thomas".[44]

By midyear, German trade unions are banned; all political parties except the Nazi Party are banned. Communist leaders remain deeply upset with Reich's emphasis upon sexual issues and problems in the family, which threaten to supplant Marxist doctrines on class struggle and the claimed great psychological health of the "New Soviet Man". Reich is expelled from the KPD, though it had functionally ceased to exist after Hitler's takeover. Having fled to Denmark, Reich's *Mass Psychology of Fascism* is soon attacked in *Arbejderbladet (The Daily Worker),* a communist newspaper in Copenhagen.[45] The Norwegian Communists claim to barr Reich from membership, though he has expressed no interests to join them. As detailed by Reich's associate Ellen Siersted,[46] they viewed Reich's *Mass Psychology of Fasicsm* as:

"...a back-stab at the [Communist] Party's existing politics 'with a cowardliness that seems to be the author's most prominent characteristic... he tries in the book to obscure the real target of his attack on revolutionary politics.' ..by stressing capitalistic society's repression of sexuality [Reich] mixed psychology into the political struggle." [46]

Numerous articles attacking Reich soon appear in newspapers in Denmark, complaining about Reich's presence. Headlines include: *"German Psychoanalyst Has Been Practicing Here and Took Payment for His Consultations"; "Wilhelm Reich, MD, denied permission to stay in Denmark"; "Psychoanalysis Outside the Three-Mile Limit".*

December: Reich is secretly expelled from the IPA, betrayed by Freud and other psychoanalysts. Jewish members of the *German Psychoanalytic Society* (DPG) are also secretly expelled. After becoming *"judenrein"* (cleansed of Jews), the DPG merges a few years later with the Nazi-run *Göring Institute*, which condemns "Jewish psychology" and endorses Hitler's *Mein Kampf.*[47]

1934: October: Reich departs Denmark for Norway. In spite of the IPA expulsion, he is invited by the Norwegian psychoanalytic group to lecture and practice his new therapy methods. He is also invited to continue with his biophysical experiments at the *University of Oslo, Psychology Department.* Reich begins publishing a new scientific-political journal, *Zeitschrift für Politische Psychologie und Sexualökonomie (ZPPS)*[6], edited under the pseudonym of "Ernst Parell." He continues to develop and argue for his theory of *Sex-Economy,* towards a more rational and less violent society through adoption of his SexPol platform, as with legal guarantees for a greater sexual freedom, and through *"self-regulation of primary natural drives in their distinction from secondary perverted drives",* the latter arising from repression of the former.

BAND: 1   HEFT: 1   1934

ZEITSCHRIFT FÜR
POLITISCHE PSYCHOLOGIE
UND SEXUALÖKONOMIE
HERAUSGEBER: E. PARELL

Reich's Research in Oslo, Norway: 1934-1939

Reich undertook significant breakthrough research in basic biology at the *University of Oslo Psychology Department* (shown below), by invitation from the Department Chairman and leader of the Norwegian psychoanalysts, Harald Schjelderup. This included the first-ever bioelectrical measurements of human sexual and emotional excitation. His later Oslo work included a microbiological study of the protoplasmic streamings in ameba and other microorganisms. This line of investigation eventually led to the discovery of the orgone energy.

INSTITUT FÜR
SEXUALÖKONOMISCHE LEBENSFORSCHUNG
BIOLOGISCHES LABORATORIUM

Reich publishes 13 papers in the first year of *ZPPS*, including *"The Orgasm as an Electrophysiological Discharge"*,[48] on the bioelectrical aspects of sexual and emotional excitation and gratification. He also publishes *"The Basic Antithesis of Vegetative Life Functions"*,[49] describing the opposing nature of the sympathetic and parasympathetic nervous systems. In these, Reich speaks more about *pulsatory functions*, of expansion versus contraction, anxiety versus pleasure, and biological motion *towards the world* in need-gratification, versus *away from the world* in pain or anxiety. He describes clinically the specific *orgasm reflex*, and a *4-phase formula for orgastic release* of inner bioelectric and colloidal-mechanical tension: *tension-charge-discharge-relaxation*. Ten other professionals contribute articles to the first year of the *ZPPS*,[6] which quickly gains an international following. It carries various reports on the struggle against Nazi oppression and on the immobility of the political centrist and left-parties in the face of disaster.

Reich also publishes political papers on sex-economic issues related to political reforms: *Dialectic Materialism and Psychoanalysis*[50] and *"What is Class Consciousness?"*[51] In these papers, he points out how sexual frustrations and family unhappiness in all classes cloud the capacity for rational social criticism and productive work. Reich's new journal retains a strong emphasis upon sexual and psychological problems, and the need for greater freedom and pleasure in life. Overall, Reich's publications identify the social and political barriers to human happiness and freedom, the scheming and violence then being widely perpetrated by the Nazis, and his increasing disagreement with Communist Party policies and leaders. This message does not go well with the Party leaders, who are often criticized by name. The early issues of Reich's *ZPPS* and other publications of this period have numerous positive references to Marx and Marxism, but predominantly negative references to the Communist Party and Soviet Union. References to Marx would also become more critical. The last issue of *ZPPS* in 1938, edited by Reich's associate Sigurd Hoel, has no mention of Marx.

1934 ATTACKS: At the *IPA Conference in Lucerne*, Switzerland, Reich is informed he has been secretly expelled from membership. He is allowed to present his scheduled paper, his last with the IPA, but only as a "guest". Attending this conference are several Americans who will later attack Reich after he has fled to the USA, such as Karl Menninger (representing American psychoanalysis) and the Soviet spies Robert Brady and his future wife Mildred Edie (Brady).[52]

Hindenburg dies and Chancellor Hitler acquires all Presidential powers. Widespread arrests and murders of anti-Nazi political opposition continues in German territory. Stalin is likewise rounding up and executing thousands in one of his many terror campaigns. Nazi and Communist newspapers in Denmark and Norway continue their attacks on Reich.

On orders from Moscow, the KPD and other European CPs are creating a list of around 3000 "troublemakers". The list is then reported to Giorgi Dimitrov, head of the *Comintern (Communist International)*, who refers the 3000 to the NKVD for subsequent arrest and execution, or deportation to the gulag.[53] These individuals, who are typically anti-Nazi as well as anti-Stalinist in sentiment, or who merely question the KPD dictates, are all labeled, correctly or not, as *"Trotskyites"* or *"counter-revolutionaries, saboteurs and wreckers."* It appears likely that Reich's name is referred to Dimitrov at this time by the KPD, or by the Soviet spy monitoring Reich, Arnold Deutsch, as previously mentioned. By 1936 Reich's name will appear on a formal NKVD death-list, along with one of his associates.

The Moscow show-trials are soon to begin, where any hint of dissent or independence from Party ideology or Moscow's dictates would result in arrest and a likely death sentence.

Reich's former IPA associate and friend, Otto Fenichel, begins writing his *Rundbrief* letters,[54] circulated privately to the refugee psychoanalytic diaspora community. Reich is periodically slandered in them, with defamatory accusations which later reemerge in the American press. Reich is one of the few who publicly writes a protest against the subordination of German psychoanalysis to Nazi agendas.[47]

1935: Reich publishes a new edition of his *Invasion of Compulsory Sexual Morality*,[31] and other papers such as *"Overview of Sex-Economic Research"* [55] and *"The Expulsion of Wilhelm Reich from the International Psycho-analytic Association"*.[56]

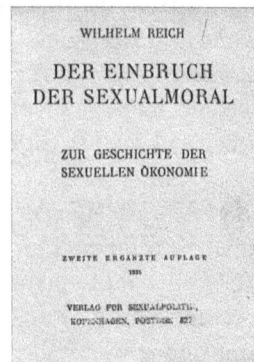

WILHELM REICH

DER EINBRUCH
DER SEXUALMORAL

ZUR GESCHICHTE DER
SEXUELLEN ÖKONOMIE

ZWEITE ERGÄNZTE AUFLAGE
1935

VERLAG FÜR SEXUALPOLITIK,
KOPENHAGEN, POSTBOKS 827

1935 ATTACKS: The Berlin Gestapo (German Secret Police) issues further orders prohibiting Reich's publications, which continue to preoccupy them:[40]

> *"Item # 41230/35 11 2 B: On the authority of the National Decree of 4/2/33, the publications 'What Is Class Consciousness?' by Ernst Parell [pseud. for W.R.] and 'Dialectical Materialism and Psychoanalysis' by Wilhelm Reich, numbers 1 & 2 of the politico-psychological series of the Sexual Political Press, Copenhagen/Prague/Zurich, together with all other publications appearing in the same series, are hereby officially impounded and withdrawn, since they are liable to endanger public security and order." Deutsches Reichsgesetzblatt,* 13 April and 7 May, 1935. [57]

Nazis enact the *Nuremberg Race Laws*, stripping Jews, Roma, Slavs and other minorities of all rights.

A half-million Comintern members are expelled in continuing Stalinist purges. The Soviet death penalty is expanded to include "spies and parasites", and anyone with knowledge of such anti-Moscow activities. By strict control of borders and of the press, and with help from Western media communists, they continue to cover up and deny such events, the Soviet Union still being misrepresented as a "worker's paradise".

1936: Reich publishes *"Psychic Contact and Vegetative Streaming"*,[58] § describing the existence of a sensible, flowing bioenergetic current in the human body, which soon he will objectively measure. He presents a unified theory of living matter, integrating contemporary works on biochemistry and bioelectricity. Reich also publishes *Sexuality in the Cultural Struggle*,[59] an early version of a later English work *The Sexual Revolution*,

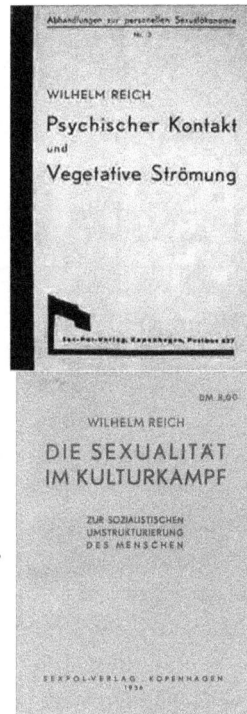

WILHELM REICH
Psychischer Kontakt
und
Vegetative Strömung

WILHELM REICH
DIE SEXUALITÄT
IM KULTURKAMPF
ZUR SOZIALISTISCHEN
UMSTRUKTURIERUNG
DES MENSCHEN

---

§ The term "psychic" as used in medical or psychoanalytic contexts, references mental or brain functions, and not parapsychology or ESP.

In Defense of Wilhelm Reich

Reich's SexPol Views 1936: *Still Relevant Today!*

"What is Sexual Chaos?

- *Pointing to the law on 'matrimonial duty' in the marriage bed.*
- *Entering into a lifelong sexual relationship without any previous sexual knowledge of the partner.*
- *'Sleeping' with a working girl because 'she's only good for that' while at the same time not asking for 'something like that' from a 'respectable' girl.*
- *Expecting a glorious wedding night after a sordid life of frequenting prostitutes, or abstinence before marriage.*
- *Regarding the act of deflowering as the pinnacle of male potency.*
- *Mentally pawing images of half naked people lewdly up and down at fourteen and then, at twenty, to become a moralist lecturing on the 'purity and honor of women'.*
- *Making it possible for someone like Julius Streicher (a Nazi propagandist) to indoctrinate thousands of young people with his perverse fantasies.*
- *To punish children for masturbation and make adolescents think that ejaculation causes them to lose spinal marrow.*
- *Tolerating the pornography industry.*
- *Making money through exciting adolescents with erotic films, but refusing them natural love and sexual gratification by reference to cultural traditions.* (continued next page)

---

exposing and criticizing the widespread sexual pathology and chaos in society, the unhappy marriages, abandoned women and children, the widespread sexual neuroses associated with guilt about "sexual sin", and with a sharp criticism of prevailing church policies and political power. Reich affirms the existence of a healthy genitality, and calls for legal contraception, public help to abandoned mothers and children, and better living and working conditions. Reich's *ZPPS* journal publishes further clarifications on *"The Cultural-Political Standpoint of SexPol"*,[60] in plain language.

Reich meets A.S. Neill, headmaster of the

> **What Sexual Chaos is Not!**
> - *To desire a sexual life with mutual love, without regard to existing laws or moral demands, and acting accordingly.*
> - *Freeing children and adolescents from feelings of sexual guilt and letting them live consistently with the aspirations of their age.*
> - *Refusing to marry or enter into a binding relationship without significant intimate knowledge of the partner.*
> - *Not bringing children into the world until one surely wants them and can bring them up.*
> - *Not to demand love or sexual favors from others.*
> - *Not to murder the partner out of jealousy.*
> - *Not having relations with prostitutes, but with friends from one's own peer group.*
> - *Not making love in alleys like the adolescents of our society, but desiring to make love in clean rooms without being disturbed.*
> - *Refusing to maintain an unhappy crushing marriage because of moral considerations, etc."*
> From *The Cultural-Political Standpoint of SexPol.*[60]

British experimental and non-compulsory *Summerhill School*, run by direct democratic assemblies of both students and teachers. This marks the start of a long friendship and professional collaboration.[61]

Reich's microscopical observations also begin, firstly as a study of the interior protoplasmic streaming motions of ameba. He notes how these cellular flowing motions are similar to the motions of smooth muscle tissue, or to whole-organism expansion and contraction. These are all, Reich argues, created through pulsatory movement of bioenergetic currents. A collaboration also begins between Reich and the French biologist Roger duTeil.

1936 ATTACKS: Reich and one of his assistants, Otto Knobel, are formally named on an NKVD death-list.[53] Marked *"Top Secret"*, the document includes Reich's name along with multiple others as "Trotskyites". The document details the arrest and execution, or deportation to the Soviet gulag labor-death camps of many persons on the list. Reich exchanged letters and possibly met with Trotsky when both lived as refugees in Scandinavia, but he was no

Top Secret Soviet NKVD Death-List, 1936 (English Translation) identifying "Trotskyites" being sought, or already arrested, executed, sent to labor/death camps, or turned over to the Gestapo. Wilhelm Reich's name appears several times.[53]

[. . .]
4 September 1936.

Top secret.

1936
To c. Sergeyev
Carry out and report
on the results of the verification of
the German émigrés.

9. 36. GD [Georgi Dimitrov][34]

To: Com. Dimitrov.
Com. Manuilsky.
Com. Moskvin.

The Cadres Department is sending to you a memorandum, "On Trotskyists and other hostile elements in the émigré community of the German CP."

We here mention only the most typical cases. Actually, the work on exposure is in progress, and a much larger number of these kinds of elements has already been revealed.

Chernomordik
Chernomordik.

[. . .]
2 September 1936.

Top secret.

MEMORANDUM

ON TROTSKYISTS AND OTHER HOSTILE ELEMENTS IN THE ÉMIGRÉ
COMMUNITY OF THE GERMAN CP

Among the German émigrés in the USSR there are people who were known
in the CPG as active Trotskyists and factionalists before their arrival
in the USSR:

[. . .]

10. OTTO KNOBEL (OTTO BRANT)—a former worker of the YCL of Germany who worked in the International Publishing House for Youth in Germany and who, in 1933, emigrated to Paris without party consent. [He] was not granted political émigré status, returned to Germany, and then emigrated again. In Paris, [he] was involved with the Trotskyists. [He] moved to Copenhagen to work, where he was not connected to the party and did not establish connections with the émigré community. In Copenhagen, [he] worked in the publishing house of Wilhelm Reich, who had been expelled from the CPG for Trotskyism. According to him, he broke with Reich over personal differences. However, according to some party comrades, he went to Berlin with Reich's consent and, a month later, in late 1935, came to the USSR via Inturist without connections and without party permission. He was so close to Reich that he read [Reich's] letters to Trotsky and even mailed them himself.

Reich's Bioelectrical Experiments[48,49,58,62]

- Makes the first-ever bioelectrical measures of human sexual and emotional excitation, in a laboratory room at the University of Oslo, Psychology Department.
- Erogenous skin surfaces (lips, tongue, nipples, genitals) have a generally higher electrical potential as compared to other, non-erogenous skin surfaces.
- Electrical potentials yield variations linked to emotions, sensation and cognitive state. The greater the intensity in subjective feeling, the greater the electrical potential. Subjective feelings of "low energy" yield low potentials, and vice-versa.
- Relaxed subjects with deep, uninhibited respiration show higher bioelectric charge as compared to "up-tight" subjects of weak or inhibited respiration.
- Pleasure reactions and "warm glowing skin" yield higher electrical potentials with sweeping pulsations. Sexual excitation and potentials are increased by gentle stroking, but reduced by aggressive or forced behavior.
- Unpleasure or annoyance reactions, of "sweaty-clammy skin" or "gooseflesh", reflect lower electrical potentials. Disappointment or fright yields a decrease in bioelectric potentials.
- The experiments indicate that the older concepts of *libido* and *vegetative currents* are a real bioenergy, which moves through parasympathetic versus sympathetic innervations into various organ systems, yielding subjective experiences of pleasure versus unpleasure or anxiety. Pleasure and fear are the same excitation moving in opposite directions; toward the core in fear and toward the periphery in pleasure.
- A better understanding of full orgastic discharge versus weaker sexual climax or inorgasmia, comes from these studies.
- Later bioelectric studies by others on human sexuality provide added detail, but largely ignore these emotional parameters, and rarely give Reich credit for his scientific priority. While modern science recognizes the relationship of the sympathetic nervous system to the fight or flight response, and the para-sympathetic nervous system to the rest and digest response, it still fails to link the emotion of fear to the former, and pleasure – especially sexual pleasure – to the latter.

In Defense of Wilhelm Reich

"Trotskyite". Large numbers of Red Army officers are now being arrested and executed for this same "Trotskyite" accusation, true or not. Reich's assistant Knobel is eventually arrested by the NKVD and sent off to the Soviet gulag, never to be heard from again. Knobel's major "crime" is his contact with Wilhelm Reich.

The non-Jewish German psychoanalysts (DPG) fully capitulate to Nazism, merging their Society into the *Göring Institute*.[47]

1936-1939: German troops occupy the Rhineland, Austria, and Czechoslovakia. Mussolini invades Libya and Ethiopia. Spain is betrayed as Stalin sells the Republican forces inferior weapons, demanding payment in gold. Stalin then orders Comintern forces to attack their one-time "allies" in the non-Soviet-controlled Republican brigades. This betrayal opens the doors to victory for Franco's Moroccan and Nazi-supported divisions.

1937: Reich publishes a new series of *Clinical and Experimental Reports*, which includes papers entitled *The Bioelectric Function Of Sexuality And Anxiety; Experimental Investigation of the Electrical Function of Sexuality and Anxiety,* and *Orgasm Reflex, Muscle Posture and Body Language*.[62] These papers describe his laboratory studies on human sexuality, including bioelectrical measurements of emotional and sexual excitation. He discovers a functional polarity between pleasurable and anxious experiences, and proposes the old Freudian libido is in reality a bioelectrical and life-energetic phenomenon. He is making experiments with his newly-discovered bions, treating cancer mice with special sterilized soil bion injections. Reich's ZPPS journal, meanwhile, continues with open discussions of his sexual theories, including criticism of the Nazis, psychoanalysis, and Marxist political parties.

Stephan Lackner, a German writer sympathetic to SexPol, argues that Reich will soon abandon Marxism, as a socially-repressive doctrine. Reich does not oppose Lackner's views.[64]

"I don't think that Reich, who has developed within the last few years as one of the most anti-dogmatic free thinkers, will

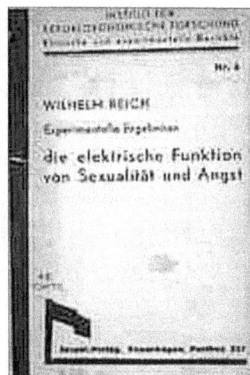

call himself a Marxist for long... [he did] adopt a system, without critical consideration, which is incompatible with Reich's own endeavors. Reich's highest goal is the liberation of the enslaved human, the freeing of their initiative. What does he hope or expect from a Communist-organized economy? It is to be seriously doubted, however, if the desired freedom of life will be possible within the rigid framework of absolute central planning, which also intrudes into the private life.[64]

1937 ATTACKS: More attacks and smears appear in the Scandinavian press. *Aftenposten* (Norway) denounces Reich's bion experiments and calls his therapy "quackery".[63]

1938: Reich publishes: *Die Bione (The Bion: Origins of Vegetative Life)*,[65] covering various experiments on his newly-discovered bion vesicle, and the observable process of bionous decay. Bions are a transitory vesicle between living and nonliving matter, created by the disintegration and breakdown of both organic and inorganic materials. They can be created by allowing matter to swell and disintegrate in water, and are speeded in this process by incandescent heating, or by freezing and thawing. Bions can then, under the correct biochemistry, bridge the gap into living forms, including via aggregate heaps into protozoans. Prof. Roger duTeil presents his paper *"Three Series of Experiments Based on the Tension-Charge Principle"*,[66] a confirmation of the bions, to the *French Academy of Sciences*.

The American physician Theodore Wolfe visits Reich in Oslo for training, and suggests he relocate to America. Reich applies for an American visa.

1938 ATTACKS: Multiple malicious smear articles attacking Reich appear in the Norwegian newspapers, especially those controlled by the Nazi and Communist parties. They attack Reich's sexual research, his new vegetotherapy, his bioelectric experiments, and his bion-biogenesis experiments. They call for his expulsion or arrest and internment in a concentration camp. The Headlines blare: *"Psychoanalytic Quackery"*, *"God Reich"*, *"Sensational*

41

Scientific Experiments". Reich is at risk of deportation, but to where? Efforts are made by Reich and his associates to refute these articles. Malinowski writes to the Norwegian press in defense of Reich, stating *"I regard his sociological works to be a distinct and valuable contribution toward science"*.[67] A.S. Neill, founder of Summerhill, writes: *"the campaign against Reich seems largely ignorant and uncivilized, more like fascism than democracy..."* [68] These efforts to defend Reich have little influence. Some of Reich's microphotos are stolen during a break-in at a photo-processing laboratory. Roger duTeil, in France, is temporarily suspended from his university post for association with Reich.

The German military mobilizes as Nazi troops march into Austria, the "Anschluss" union, and invade the Sudetenland of Czechoslovakia. At a September "peace conference" in Munich, Hitler swears he is now satisfied and wants no more land. Labor Party PM Chamberlain of Britain returns home to smile before world cameras, waving a meaningless scrap of paper with Hitler's signature, proclaiming *"Peace for our time"*. *Kristallnacht* takes place on 9-10 November; Jews are beaten up, murdered, their businesses smashed, synagogues burned.

1939: In January, Reich develops a special *sand-packet* (SAPA) bion culture, made from beach sand heated to incandescence. The SAPA bions generate a strong visible and sensible bluish foggy radiation. Over subsequent months, Reich finds this radiation will register on camera film plates, and yield anomalous electrostatic and magnetic phenomena. He develops special fluorescing screens and Faraday-type enclosures to amplify the phenomenon. By those and other experiments, he determines this to be a new form of energy, a life-energy, and it is given the name *orgone energy*. This discovery leads directly to the invention of the *orgone energy accumulator*, an enclosure made to capture and isolate the phenomenon. The orgone energy is eventually clarified as having both organismic and atmospheric expressions, also filling the atmosphere as an energetic continuum.

Reich publishes *"Bion Experiments on the Cancer Problem"*[70] and *"Three Experiments with the Static Electroscope"*[71] detailing his new ideas and findings. Neill writes favorably about Reich in several of his books, such as *The Problem Teacher*[72] and later *The Problem Family*,[72] exposing hidden cruelty to children, and bringing

Reich's Bion Observations and Discoveries
65,66,69,70,138

- Demands use of higher magnifications and study of living preparations, without the "dead" fixing or staining of slides.
- Originally a study of streaming motions within moving and expanding-contracting ameba, as a corollary to his bioelectrical experiments, which measured streaming sensations and similar expansion-contraction in the human organism. Ameba *"reach out towards the world"* to acquire food (meeting of needs) or contract into a ball, *"away from the world"* when threatened.
- Organic materials in water will swell and disintegrate into small ovoid bion vesicles, about 1 micron diameter. Disintegration and bion-forming processes are speeded by boiling, heating to incandescence prior to immersion, and by freezing/thawing. Bions exhibit qualities of both the living and non-living.
- Bion vesicles, once developed, yield increasingly life-like motions, depending upon their parent materials and the chemistry of their solutions. Some bion preparations, derived from and kept within sterile solutions, will replicate.
- Bion vesicles can aggregate to form larger microbial entities, such as ameba, vorticella, or paramecia.
- Reich notes a parallel between the organization of bions from rotting organic material in pond water, into ameboid forms, to the disintegration of human body tissues in the cancer process, where ameboid cancer cells are developed.
- A special beach-sand bion preparation is discovered, yielding blueglowing bions with radiant energy properties.
- A new *Reich Blood Test* is developed to diagnose the tendency towards cancer, years before the appearance of tumors.
- Reich's discoveries constitute a solution to the riddles of the *origins of life,* as well as the *origins of the cancer cell.*
- These findings pre-date and anticipate much of modern astrobiology and biogenesis research, the discovery of deep-ocean geothermal vents thick with life-forms, the findings on Archaea and extremophile microbes, all of which are bionous in both appearance and development.
- Reich is rarely given credit for his scientific priority on the bions, even while being verified through different terms and theories. The modern term of *apoptosis* reflects Reich's discovery of the *bionous disintegration* of cells, for example.

Modern Replications of Reich's Bion Discovery

Reich's critics honk and sneer, *"Only Reichians can see the bions"*. But bions are quite real, with many published bion photographs. One such microphoto is given below, taken by the author at his laboratory where seminars are held. Below that is a screen-shot of a YouTube video on bions, from a presentation by the author to the *Second International Conference on Pleomorphism in Health and Disease*,[69] where the reader can get additional images and citations.

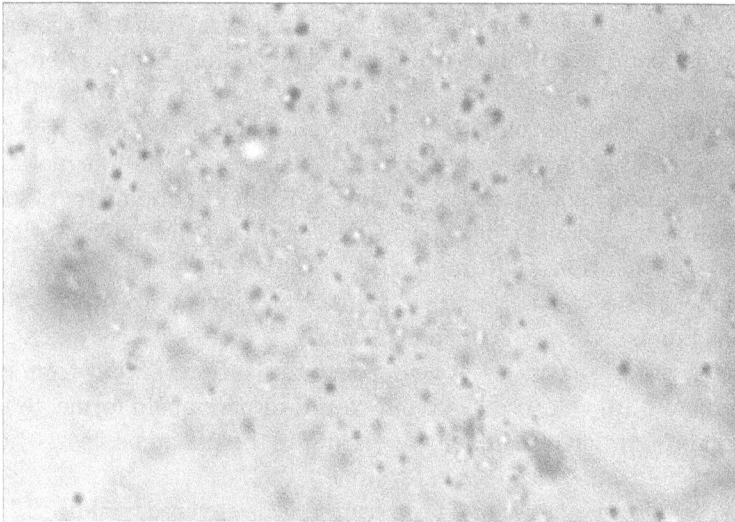

WILHELM REICH'S BION-BIOGENESIS DISCOVERIES - E...

http://youtu.be/-PVnS72IIY8

Subjective Anomalies in the Orgone Accumulator

While working in his Oslo laboratory, Reich noted a strong radiation coming off of special bion preparations made from beach sand (SAPA). The radiation could be felt on the skin, and it triggered conjunctivitis when viewed in the microscope for extended periods. The laboratory room holding multiple cultures of the SAPA bions became sensibly charged when the preparations were stored in metal cabinets. Anomalous electrostatic and magnetic effects appeared in the lab. When viewed in the dark, the room was filled with foggy forms and scintillating light-points. Reich reported a skin-tanning effect from the SAPA radiation, during winter and right through his clothing. SAPA bions were injected into cancer-mice, reducing their tumors. He puzzled over these effects, which ultimately led to the creation of a metal box with viewing lens, surrounded by dielectric organic material, by which to study the unusual radiation. He objectified the radiation from bions, but also a free form in the open air. The *orgone energy* and *orgone accumulator* thereby came from the same basic set of observations and empirical discovery.

Orgone Accumulator Charger with Eyepiece
at the author's laboratory.

In Defense of Wilhelm Reich

Reich's work to a new public in the English-speaking world. He also writes on the subject of *Work-Democracy*,[73] producing the first in a series of papers over the next six years on the subject of the natural and self-regulated organization of work functions.[68]

1939 ATTACKS: March 15: Nazis invade the remainder of Czechoslovakia, in violation of prior agreements at Munich.

August 19: Constant and growing Nazi power, attacks and betrayals by Communists and psychoanalysts, complete with public slander and alarm in newspapers, force Reich into flight once again, this time headed to New York City in America. He departs on the *Stavenger Fjord*, the last ship leaving Norway for America just before the outbreak of WW-II. Elsa Lindenberg does not follow, however, in spite of Reich's pleas she do so.

August 23: The Soviet Union and Germany sign a mutual nonaggression treaty, the *Hitler-Stalin Pact*. Marxists globally are initially shocked at the "unbelievable" news. Nearly all were unaware of the Soviet Union's secret cooperations and rearmament deals with the German High Command, and later with the Nazis, going back more than a decade.[3] But the majority of leftists blindly obey their leaders and accept their marching orders, *to now defend Hitler and the Nazis as "fellow socialists"*. This Pact carries secret side agreements on the division of Europe between the Nazis and Soviets following their planned invasions. The Nazis also sign the *Pact of Steel* with Mussolini. Britain and France warn Hitler they will not accept any further incursions, but after their appeasement in Munich, they are ignored. War is imminent.

September 1: The German Wehrmacht (Army) invades Western Poland, the Luftwaffe (Air Force) attacks civilian and military targets. Warsaw is bombed and shelled into ruins. Two weeks later, as per the secret side-agreements of the Hitler-Stalin Pact, the Soviets invade Eastern Poland, which is swallowed up by Communist tyranny and slaughter. Stalin's Red Army invades the Baltic states of Latvia, Lithuania and Estonia, forcing them to accept Soviet troops and military bases on their territory, a prelude to forced annexation.

September 23: Freud dies in his London exile, having fled Vienna shortly after the *Anschluss* (Nazi annexation) one year previously.

November 30: The Soviets invade Finland.

The Norwegian ship *Stavenger Fjord*, which took Reich to America on 19 August 1939, the last ship to leave before the outbreak of World War II. Getting out, just in time.

The *Hitler-Stalin Pact*, as viewed from the world's democracies. Both dictators plotted together against the world, in alliance with Imperial Japan and Fascist Italy, and against each other, empowered by millions of violent, sexually-disturbed followers.

In Defense of Wilhelm Reich

WW-II intensifies as various cities and nations are swallowed whole by the Nazi military machine, or by the Red Army. Various regional Communist Parties across Europe, under orders from Moscow, welcome the invading Nazis as "comrades", even as they goose-step and shoot their way across Europe. Stalin continues with Communist Party purges and show-trials, massacring former "comrades" who had themselves killed other "comrades" and civilians in earlier atrocities. "Non-comrades" continue to be deported by the millions into the Soviet gulag death-camps. Frequently these "counter-revolutionary wreckers" are entire villages of ordinary men, women and children of all ages, peasant farmers rousted from their homes at gunpoint in the middle of winter, and quickly forced into railroad cars for a long trip to Siberian slave-labor-death camps, with little more than the clothing on their backs. Red Army and NKVD officers then confiscate their property, take up residence in their homes, and divide the spoils among their cadres. Women are violated, drunken looting prevails, protestors or those who fight back are summarily shot, in what is little more than bands of well-armed "comrade" criminals lording over a disarmed and cowed population. Nazi troops, while more "disciplined and orderly" in their treatment of conquered civilians, are unleashing their own deadly plots and atrocities. It will take many decades after the end of the 2nd World War before the full extent of Nazi and Soviet atrocities are exposed to the world

Reich's American Period

While Reich's escape to the United States removed him safely beyond the reach of the Nazis, he was stepping into a beehive of communist activity, including Soviet sympathizers and spies of various sorts who operated relatively freely within America. The American social and political landscape had become something of a mine-field for unorthodox physicians and anti-Soviet liberals escaping from Europe, such as Reich, irrespective of their strong anti-Nazi views. Many of his old enemies in psychoanalysis had also emigrated to the USA, where they would continue to attack Reich with malicious slander. I will therefore backtrack a bit in time, to describe key events that would affect Reich within America, and the people involved in those events.

1910-1940s Prelude to Attacks: In 1910, Abraham Flexner, working on behalf of the *American Medical Association* (AMA), published his infamous *Medical Education in the United States and Canada*,[75] calling for dramatic changes in health care teaching institutions and law. The *Flexner Report* disparaged herbalism, midwifery, women and black physicians, naturopathy, homeopathy, chiropractic and osteopathy, calling their practitioners "quacks" and advocating a closure of their teaching institutions. Under guidance from the Flexner Report and financed by the *Carnegie and Rockefeller Foundations*, new laws were instituted over the next 30 years in the various states to unscientifically elevate the practitioners of allopathic medicine and the MD degree, and criminalize all other healing methods.[76]

1920-1940s Prelude to Attacks: Roger Baldwin and around 60 other progressive-socialist-communist and pro-Soviet political activists, joined to form the *American Civil Liberties Union* (ACLU). These founding members included Arthur Garfield Hays, who in the 1940s became Reich's personal attorney, and gave him compromised advice at a critical time. Another ACLU founder, Felix Frankfurter, a Harvard law professor and later US Supreme Court judge, would later issue alarming and arrogant legal opinions that were central to the denial of Reich's legal appeals, and hence, to his imprisonment and death, and to the banning and burning of his books.

Baldwin was never involved in the persecution and death of Reich, so far as is known, but his opinions as head of the ACLU carried considerable weight among American liberals, including other ACLU founders like Hays and Frankfurter. They were tutored on Baldwin's ideas and activism, and their biographies eventually crossed paths with that of Wilhelm Reich, to no good end.

Baldwin wrote effusive praise for the Soviet Union in his 1928 book *Liberty Under the Soviets*, white-washing their crimes, and he steadfastly held to his views until the end of his life. In a 1934 article for the Comintern publication *Soviet Russia Today*, he wrote:

> ***Proletarian Liberty in Practice:*** *When that power of the working class is once achieved, as it has been only in the Soviet Union, I am for maintaining it by any means whatever. Dictatorship is the obvious means in a world of enemies at home and abroad. ... [I]f American champions of civil liberty*

could all think in terms of economic freedom as the goal of their labors, they too would accept 'workers democracy' as far superior to what the capitalist world offers to any but a small minority. Yes, and they would accept – regretfully, of course – the necessity of dictatorship while the job of reorganizing society on a socialist basis is being done."[77]

In 1950, retiring from the ACLU, Baldwin stiffly reaffirmed:

"I am for socialism, disarmament, and ultimately, for abolishing the state itself as an instrument of violence and compulsion. I seek social ownership of property, the abolition of the properties class, and sole control of those who produce wealth. Communism is the goal. ... I knew what I was doing. I was not an innocent liberal. I wanted what the communists wanted and I traveled the United Front road to get it."[78]

When speaking of Baldwin, Arthur Hays admiringly declared, *"The ACLU is Roger Baldwin"*.[79] Hays' left-wing credentials were further certified in his many connections with different Communist and pro-Soviet front groups. He became Chief Counsel for the ACLU, and was also a founding member of the *All-America Anti-Imperialist League,* the *American Society for Cultural Relations with Russia,* the *Victor Berger National Foundation* (honoring an American communist revolutionary convicted of sedition), the *Christian Social Action Movement* (to bring socialism-communism into American churches), the *Committee on Coal and Giant Power* (attempting to nationalize the US coal industry), the *John Reed Club* (honoring the "first American Communist" and working to oppose public criticism of communism), and the *Mooney-Billings Committee* (to release convicted Stalinist terrorists who killed 10 people with a bomb).[80]

Hays frequently spoke in favor of the Communist Party and against anti-communism on American radio.[81] He was a traveller to the Soviet Union as early as 1926,[82] and an observer at the later Moscow "show trials", when only those with CP approvals could travel freely in the USSR. He also worked as estate attorney for the Eitingon family's lucrative Russian-Soviet fur trading business, a family which included Soviet spies and a KGB assassin.[83]

At the behest of American communists, in 1927 Hays and Felix

Frankfurter served as co-counsels at the murder trial of anarchists Sacco and Vanzetti. Hays was also asked (along with Frankfurter) to act as defense attorney at the 1933 trial for five Comintern members accused of starting the Reichstag fire in Germany. Hays helped get four of the accused released, including one Giorgi Dimitrov, who returned to Russia a hero. He was quickly appointed to lead the Soviet Comintern and became a confidant of Stalin. In that role, Dimitrov referred Reich and thousands of other dissenters to the NKVD,[84] for arrest and almost certain death should they be caught.

In one early 1951 TV broadcast, Hays defended American communists with great passion, declaring the anti-communist Senator Joseph McCarthy to be *"...the most dangerous man in the United States. ...more dangerous to freedom than all the communists in this country"*.[85] Especially by the 1950s, there was abundant public evidence on Soviet totalitarianism, such as the show trials, the gulag, Ukrainian genocide, and other massive slaughters which had taken place at the hands of fanatical communists internationally. Many Soviet agents had been exposed by the various House and Senate committees (including Alger Hiss, the Rosenbergs and others), with nuclear-armed Soviet totalitarianism on the march globally. But Hays nevertheless proclaimed American *anti*-communism to be far more dangerous. Hays was also an advisor to President Franklin D. Roosevelt, where such pro-Soviet ideas pushed America into deadly appeasement of Stalin.

1935-1938 Prelude to Attacks: Arthur Hays, Robert Brady and Mildred Edie Brady were also co-founding members of the *Consumer's Union* (CU),[52,86] a Stalinist break-away organization formed in 1936 after a violent labor strike against the original anti-Stalinist *Consumer's Research* (CR) organization. For its first 20 years, CU was led by Arthur Kallet, a communist radical and staff writer for the Soviet-controlled *Daily Worker*. Kallet also supported the Hitler-Stalin Pact.[87]

Mildred Edie Brady led the *CU Western Council* during its subversive years, later rising to attain editorship over its primary publication *Consumers' Reports*. She also played a key role, through CU lobbying of Congress to shape FDA policies and empower them as a Soviet-style social police force, complete with spy networks, a "Ministry of Truth" lording over speech and press, and heavily armed FDA enforcers. Her husband Robert, an economics professor

In Defense of Wilhelm Reich

and CU officer, appeared as a guiding force within many communist-front organizations. He belonged to many of the same communist groups as Hays, travelled freely to the Soviet Union in 1935, and together with his wife Mildred was a close personal friend and operative within one of the most effective Soviet spy rings run in America, led by Robert Brady's former student, Nathan Gregory Silvermaster.[52] Another of Brady's spy-students was Norman Bursler. Both Silvermaster and Bursler were confirmed as Soviet agents years later in the *Soviet Venona decryptions*. Mildred Brady would later work within another effective and dangerous Soviet spy group, led by KGB moles Michael Whitney Straight (of the Cambridge spy ring) and Henry Wallace, in their joint controls over the *New Republic* magazine which published Brady's 1947 slanders of Reich.[52]

Robert Brady's 1937 book *The Spirit and Structure of German Fascism*[88] slandered free-market capitalism as being no different from Nazism or Mussolini Fascism, and issued a call-to-arms for *"the workers"* towards *"...the **extermination** of the rich and powerful of bank, factory, bivouac and cloister"* (emphasis added, JD).[88] His book was written before the Hitler-Stalin Pact of 24 August 1939, but that Pact did not deter Brady, as he signed on to an *"Open Letter in Support of the Soviet Union"* published in the glossy Comintern magazine *Soviet Russia Today* in September of that same year.[89]

One year earlier, Mildred Brady, in her role as Soviet spy and head of the Western CU, recruited Robert Oppenheimer, key scientist in the American atomic bomb program, as one of her "consumer activist" cadres, a point of contact which would later figure into Oppenheimer's loss of Top Secret security clearances.[90] In a letter to Brady dated 29 May 1938, he addressed her with some affection as *"Dear Boss"*, and organized a meet-up.[91] The Bradys played key roles in numerous Soviet organizations, including CU, even as they maintained positions within the Roosevelt *Office of Price Administration*. They were exposed as Soviet agents in 1941, losing those jobs.[92] Mildred Brady as well as Oppenheimer would years later independently denounce Reich.

Felix Frankfurter, another progressive-socialist-communist ACLU founder previously mentioned and who would cross paths with Reich, was also a key advisor to President Roosevelt and a member of many communist-front groups. He joined the *Griffin Bill Committee* (fighting to exempt aliens from military service), the *National Popular Government League* (a consortium of various progressive-socialist and communist organizations advocating for government ownership of private industry and utilities), and endorsed the pro-Soviet book *Professional Patriots*, which attacked anti-Communists as being motivated by personal greed and commercialism.[93] He was additionally *"good friends"*, based upon shared leftist philosophy, with Harold Laski,[94] a British Marxist economist who wrote a glowing foreword to Robert Brady's aforementioned book denouncing capitalism as equal to Nazism.[88] Frankfurter also *"frequently"* wrote unsigned articles and editorials for the Soviet-controlled *New Republic* magazine,[94] whose pages would also carry several articles attacking Wilhelm Reich. As a Roosevelt-appointed US Supreme Court justice, Frankfurter's legal opinions were later cited (along with smear articles by Mildred Brady, discussed shortly) as *legal justifications* for FDA book-burning and Reich's imprisonment.

The *US Food and Drug Administration* (FDA) was also busy in 1938 under the socialist-progressive Roosevelt Administration, when they received vast new regulatory powers and funding. Originally formed in 1906 to standardize quality and purity of food and drugs, *the 1938 laws gave the FDA police enforcement powers,* including the rights to make unannounced inspections, to prohibit free speech via suppression of therapeutic claims for drugs, cosmetics and medical devices they deemed to be untrue, and to similarly control product advertising. While some of the reasons for the new legal powers were arguably legitimate, given documented cases of medicines containing toxic compounds, the FDA rarely used its enforcement powers against the poison pills or questionable medical devices being prescribed by the AMA-approved MD-hospital doctors. Overall, the prosecutorial arm of the FDA worked closely with the AMA and pharmaceutical supply houses to attack as "quacks" the non-conventional physicians and natural health practitioners, such as midwives, herbalists, chiropractors, osteopaths and homeopaths.

The FDA also began to ban and burn publications giving advice on health and healing which went against AMA-dictated treatments. Only in a few cases – as when J.I. Rodale spent a fortune to legally fight off an FDA injunction aiming to destroy his *Organic Gardening* magazine (which promoted vitamins and foods grown on organic soils) – was the FDA prevented from carrying out its increasingly aggressive and anti-Constitutional activities. Harry Hoxsey's popular and effective cancer-treatment clinics, using Native

In Defense of Wilhelm Reich

American herbal remedies, also was attacked by the FDA and he fought back strenuously with expensive attorneys, but in the end the FDA ground him down and his popular and effective cancer treatment clinics were shuttered. Multiple therapeutic natural hot springs across America, and the inexpensive clinics located on their grounds, were also forcibly shut down by the FDA "medical police", who claimed their waters were "toxic" to the many who flocked to them for beneficial treatments of arthritis and other health problems.[76] In the public eye – due to widespread media lying about "dangerous quacks" versus "responsible MDs" (who were then endorsing cigarette smoking and all kinds of expensive toxic remedies) – the FDA came off as "heroic" and "rational". Few knew that the "advertising" they banned and burned was really scientific or medical literature.

By the time Reich arrived in the USA, he was fed up with the betrayals by the European and Soviet Communists, and held growing anticommunist sentiments. This occurred at the very time when Communists in America were typically organized under a pro-Soviet command structure, getting funds and directives from Moscow, but concealing their agendas due to a growing and quite rational public anticommunist sentiment. Ordinary Americans were properly concerned about Marxist-Leninist-Stalinist agendas, and the totalitarianism to come if they ever got power.[95] Reich's European criticisms of Marxist ideology were greeted with tremendous hostility by the KPD, which cooperated fully with the Stalinist death-machine, to the point of publicly denouncing him, expelling him from their organizations, reporting him to Moscow as a troublemaker, and getting him on an NKVD death-list. The reaction from the American Comintern would be hardly any different. The progressive-leftist Roosevelt FDA and medical infrastructure also had a well-oiled machine for crushing down new and unorthodox medical therapies and their pioneers.[75] And they were helped along in this cause by activist groups like the *Consumers' Union*, which worked steadfastly towards Soviet-style "Big Government" agendas.[52]

Continuing with Reich, 1939: August 27th, Reich arrives in America, invited to work and teach by Theodore Wolfe, a physician specializing in psychosomatic disorders. Reich establishes himself in Forest Hills, NY, with a new laboratory and therapy practice, and several new trainees. He hires Ilse Ollendorff as his lab assistant, who eventually becomes his third wife. Reich is temporarily free from the most serious and deadly threats from the Nazis and Communists who hounded him in Europe over the last decade. For the next decade, he is free to pursue his research largely unmolested. Reich also ends his involvement with political parties and direct social activism, focusing upon his new experimental and clinical discoveries. Republished versions of his papers and books from the European period typically undergo a redraft to eliminate much of the older Marxist language, which Reich now feels is unnecessary or inaccurate.[113]

1940: Reich begins teaching a course on his work at the *New School For Social Research*, NYC, a "University in Exile" for foreign scholars fleeing Hitler's Europe. Additional physicians are attracted to his work, including a few American psychoanalysts and psychiatrists. Reich resumes experimental work on cancer mice and SAPA bions, and on methods for objectifying the newly discovered biological and atmospheric orgone energy. New experiments and apparatus are constructed to objectify the visible orgone scintillations in darkrooms, inside the orgone accumulator, and in the open sky. His first experiments on the orgone accumulator's thermal and electroscopical anomalies are underway.

Reich also begins in-depth experimental study and therapy of the *cancer biopathy* in humans. He observes a loss of bioenergy charge of the tissues due to chronic sexual stasis, respiratory blocking, and emotional resignation. As tissues lose bioenergetic integrity, they break down into bions, which thereafter aggregate to form ameboid cancer cells, much as he observes in moss and grass infusions in the formation of protozoans. These, over time, can organize into tumors. His therapy can increase autonomic pulsation and respiration, and together with use of the orgone energy accumulator, will increase the vitality and charge of the body and tissues. He also develops the *Reich Blood Test* to evaluate the energetic integrity of red blood cells within stressing physiological saline solutions. He develops methods for identifying precancerous conditions and cell forms within body fluids. The clinical benefits of his new diagnostic and therapeutic approaches are rigorously studied by Reich and other physicians in his working group.[6]

In November, he writes to A.S. Neill about America, and his loss of confidence in socialism:

The Orgone Accumulator

Resembling a *Faraday Cage* or *Hollow Capacitor*, the orgone accumulator acquires a charge of a previously unknown energy directly from the atmosphere and cosmic space. The interior charge is documented in many biological experiments, as with the stimulated growth of seedlings, speeded healing of animal tissues, stimulus of the human parasympathetic nervous system and boost in immune system functions. Anomalous physical effects develop inside, such as a slightly higher temperature and an increase in electrostatic charge density. The accumulator can also increase the count-rates of certain nuclear radiation instruments, increase the spectrographic absorption of charged water in UV frequencies, and increase various "ionization" effects within high-vacuum tubes. The photo below shows two human-sized orgone accumulators inside a larger orgone energy dark-room at the author's laboratory, used for experiments.[97]

Reich's orgone energy and the accumulator are world-class discoveries, confirmed many times by independent scientists and physicians. However, their findings on the accumulator continue to be slandered and misrepresented by the irrational critics, who never reproduced the pertinent experiments, and deliberately lie about nearly everything.

"...this so-called bourgeois society has done more in the course of seven years under capitalistic rules in the field of social security than any communist in Russia would dream of getting. ... I feel myself completely confused and inclined to revise most of the things I ever learned in Europe about what socialism should be. ... If you hear from socialists and communists coming over here and claiming that Roosevelt is a dictator or a fascist, then your stomach simply turns around. I started to hate them." [96]

In December 1940, Reich begins a letter exchange with Albert Einstein, on his orgone biophysical discoveries.[98]

1941: Reich meets with Albert Einstein for five hours on 13 January, at Einstein's Princeton home, discussing his findings. In a later meeting, Reich brings and demonstrates the thermal anomaly (To-T effect) with a small orgone accumulator, and visual observations of orgone energy scintillations in the *orgonoscope* device. Einstein says Reich's discovery, if true, would be a "great bomb" for physics, and asks for loan of Reich's equipment so he can study the phenomenon. In early February, Einstein reports to Reich a positive result on his verification tests, but then claims there is a classical explanation for this, notably convection currents in the room. He also disassembles the loaned accumulator, after which it ceases to function properly. Reich rightly objects to Einstein's mechanical explanation and gives him a series of control experiments which can be undertaken to rule out the convection explanation. Einstein never does those experiments, unfortunately, and goes silent, for reasons uncertain, but possibly due to national security issues related to the atomic bomb program. Only in November does Einstein return the loaned equipment to Reich.[98]

Reich ends his teaching at the *New School for Social Research*, but continues with private research and educational efforts.

1941 Attacks: Robert and Mildred Brady are identified by Congressman Martin Dies, Chair of the *House Committee on Un-American Activities* (HUAC) for ties to the Communist Party. They are subsequently fired from their posts in the Roosevelt *Office of*

The Orgone Accumulator Thermal Anomaly (To-T)

Einstein firstly declared to Reich that the To-T effect was *"a great bomb" in physics*, but later claimed the effect was the product of air convection currents in the room. Reich encouraged him to undertake a series of control experiments which refuted the convection argument, but Einstein never did so.[98] However, others have, including the author of this book, in a very tightly controlled 6-year experiment yielding significant results.[101]

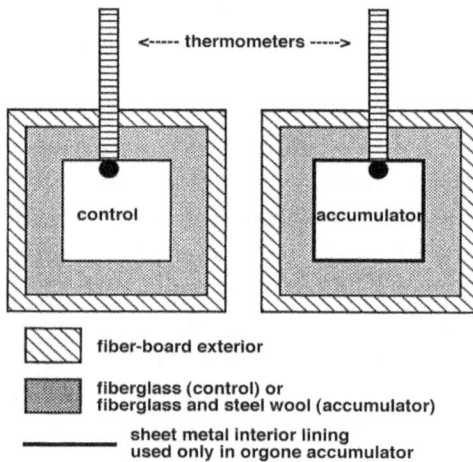

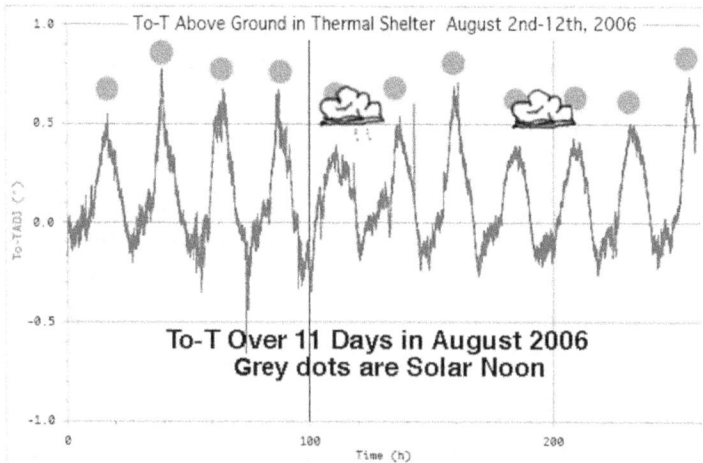

http://www.orgonelab.org/DeMeoToTSubtleEnergies.pdf

Price Administration.[92] They are then praised as heroes in the leftist *New Republic*,[99] which only five years later will print slanderous articles by Mildred Brady attacking Reich.

By mid-1941, the FBI opens a file on Wilhelm Reich, based upon anonymous and patently wrong information.[99] He is accused of having been a *"paid agent of the Austrian Communist Party"*, and a "member of the Norwegian CP" at the very time he was being publicly slandered by the Nazi and Communist press in Scandinavia. Dr. *Wilhelm* Reich is also being confused with a Dr. *William* Reich, a known activist member of the CP-USA also living in New York City. These errors are amplified when the FBI identifies his ex-wife Annie Reich-Rubinstein and her new husband as communists. By July 1941, the FBI concludes Reich should be *"considered for custodial detention in the event of a national emergency."*[100]

Late in 1941, following the Japanese attack on Pearl Harbor and Germany declaring war on America, Reich is arrested and held in detention for 3 weeks on Ellis Island, along with numerous other recent German, Japanese and Italian emigres. The FBI investigates Reich's background. Once the FBI confirms he is not *William* Reich, verifies his break with the communists in 1933, and that he is not a Nazi supporter, he is released.

1942: Reich publishes the 1st English edition of *Function of the Orgasm,*[102] with the preface *Discovery of the Orgone, Vol.1.* This provides an introduction to his European clinical and experimental findings, for the English-speaking world.

Reich also begins publication of a new journal, the *International Journal of Sex-Economy and Orgone Research (IJSO).*[103] The *IJSO* presents cutting-edge discussions on problems of child-development, family and sexual life, and social barriers to human freedom and happiness. Articles discuss his new emotion-release *orgone therapy*, which Reich had developed in Europe (previously called *vegetotherapy*), as well as the newly

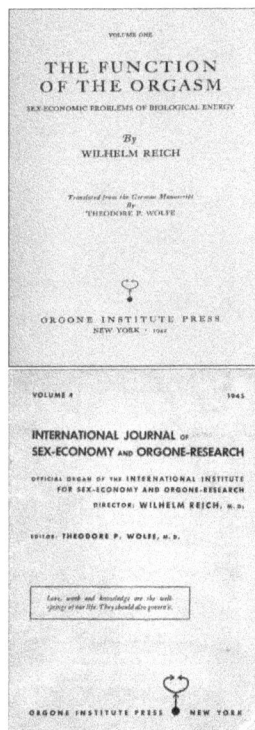

Biological Effects of the Orgone Accumulator

Reich and his associates undertook many clinical and controlled experimental studies proving the orgone radiation could benefit the growth and health of plants, laboratory mice, and humans. The studies are myriad and beyond the scope of this publication to review in any detail, but various citations are given in the Reference section. (Also see the summary in Chapter 3.) Below is a typical result from one run of a controlled 3-year mung bean seed-sprouting experiment at the author's laboratory, under optimal conditions. One group of seeds is sprouted inside the orgone accumulator, and compared to a sprouting control group kept in a non-accumulating enclosure with all other factors being the same. On average, a 34% boost! (p<0.0001)[104]

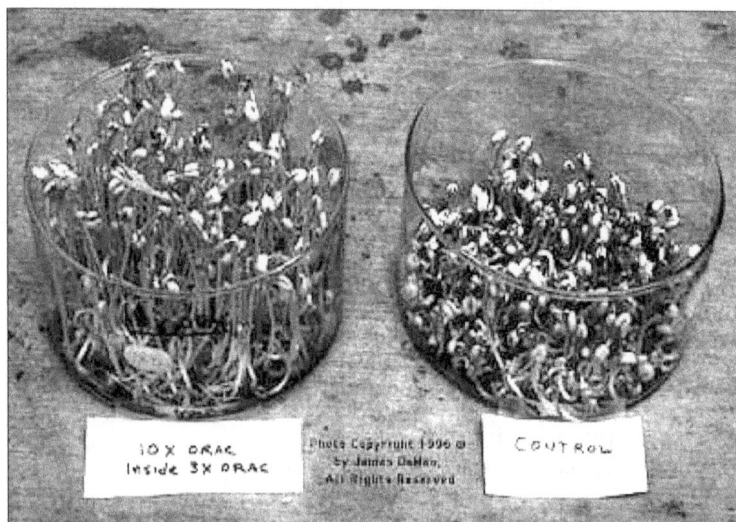

TABLE 1.	Control Groups	Orgone-Charged Groups	Percent Change
Average Seedling Lengths	149 mm	200 mm	+ 34%
Germination	95.8%	97.3%	+ 1.6%
Weight Increase	49.0 gram	53.2 gram	+ 8.6%
Average Water Consumed	109.9 ml	118.3 ml	+ 7.6%
Refractive Index (%Brix)	6.3	5.1	− 19%

http://www.orgonelab.org/DeMeoSeedsSubtleEnergies.pdf

discovered *orgone energy*. Reich has by now developed a better understanding of the cancer process, and is investigating the orgone energy accumulator for symptom-reductions in various disease biopathies. His bion experiments are refined and replicated by others. Reich also continues investigations on the principles of *orgone physics*, to better understand the orgone energy and its properties. Reich's major papers for this year include: *"Biophysical Functionalism and Mechanistic Natural Science"*, *"The Discovery of the Orgone: Experimental Investigations of Biological Energy"*, *"The Carcinomatous Shrinking Biopathy"*, and *"The Natural Organization of Protozoa from Orgone Energy Vesicles"*.[6] Reich's new journal includes 16 separate contributions by others in the first year of publication, verifying and expanding upon his findings.

1942 ATTACKS: Distorted "book reviews" of Reich's new journal, the *IJSO*, and his *Function of the Orgasm,* appear in the *American Journal of Psychiatry* and the *Journal of the American Medical Association*.[105]

Karl Menninger also spreads sexual slander against Reich:

"This journal [IJSO] is the amazing product of another one of the many secessions from psychoanalysis. This one is led by the erratic Wilhelm Reich, supported by Helen Dunbar's ex-husband [Theodore Wolfe]. To get the main idea one should read the article entitled: 'Vegetotherapy'. The patient relaxes upon the couch and learns to breathe in such a way as to bring about an orgasm. Fortunately this is not called psychoanalysis. Many fine sounding words such as 'orgone' and 'bionic' are introduced. One can nearly always find one good idea in the craziest contribution, but in this instance I failed."[106]

During the War years, Menninger's clinic in Topeka Kansas becomes a safe harbor for various Marxist psychoanalysts and Reich-haters, such as Otto Fenichel. Menninger had also represented the American branch of the IPA at the time of Reich's expulsion at the 1934 Lucerne Conference in Switzerland.

Reich and his associates write and publish rebuttals.[107]

1943: Reich purchases a new property in the mountains near Rangeley, Maine, in hopes to develop a research center and clinic.

He publishes: *"Experimental Orgone Therapy of the Cancer Biopathy"*, *"Biological Miscalculation in the Human Struggle for Freedom"*, and *"Work Democracy Versus Politics"*.[6] Reich's associates also publish 6 new papers supporting his findings.[6]

1943 ATTACKS: Reich's *Function of the Orgasm* is given a malicious Review in *Psychosomatic Medicine*.[108]

1944: Reich publishes new experimental papers: *"Thermal and Electroscopical Orgonometry"* and *"Orgonotic Pulsation: The Differentiation of Orgone Energy from Electromagnetism"*.[6] His associates publish 14 new articles either confirming his findings or defending him from recent irrational attacks.[6] By this time, Reich has invented the *orgone energy field meter*, and a special *orgonotic pulsation demonstrator,*[109] elaborating on increasingly sophisticated methods for demonstration of the orgone energy, and distinction of it from other forms of energy such as electromagnetism. Reich is also gradually shifting his residence and laboratory from New York to his new property being developed in Rangeley, Maine. These various new steps, made possible by the freer social atmosphere in America, have attracted more people to his side. He has become a harsh critic of Communism, emphasizing his own sex-economic and orgone biophysical discoveries over his prior psychoanalytic and political activities. A sign on Reich's laboratory wall reads: *Work, Not Politics!* Peter Reich is born to Ilse and Wilhelm Reich.

1944 ATTACKS: A new "whispering campaign" of slander against Reich is proceeding within the American medical-psychiatric and psychoanalytic groups. This includes a repeating of old slanders spread in prior years by European psychoanalysts who also had emigrated to the USA around the same time as Reich, to escape Nazism. The gossipers include his former friend Otto Fenichel, and ex-wife Annie Reich, who from hidden positions slander Reich with the lie that he was promoting some kind of "masturbation therapy". Reich's associates publish rebuttals, and the whisperers are threatened with legal action. They all back down or deny ever saying such things.[1,51,62]

1945: Reich publishes an English language version of *Character Analysis*[110] and *The Sexual Revolution*.[111] His new clinical and experimental papers include: *"The Masochistic Character"*, *"Anorgonia in the Carcinomatous Shrinking Biopathy"*, *"Orgone Biophysics, Mechanistic Science and 'Atomic Energy'"*, *"Experimental Demonstration of the Physical Orgone Energy"*, and *"The Development of the Authoritarian State Apparatus from Rational Social Relationships"*.[6]

1946: Reich becomes a US Citizen. He also publishes a revised English edition of his *Mass Psychology of Fascism*,[112] a work which presents deep insights into the development and rise of the Nazi movement, due to emotional-sexual repression of the masses, and the associated manipulation of people along similar mass-psychological lines. He exposes the role of the organized Church in facilitating sex-repression and the resulting syndrome of blind obedience to authority. Reich's *Mass Psychology* also contains insightful criticisms of Marxist thought, and the failures of left-political movements. Reich also deletes much of the Marxist jargon of the earlier German edition, and expands his criticisms of Soviet communism.[113] In later years, he even more strongly emphasizes the functional similarity between the Nazi and the Communist in his discussions on *Black versus Red Fascism*. These criticisms will quickly gain the attention of cloaked Stalinists working within the American public media and political institutions.

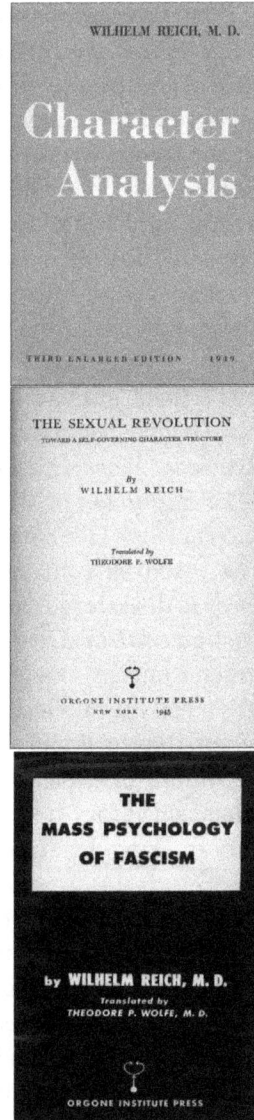

1946 ATTACKS: Reich is attacked by the Stalinist writer Fredric Wertham (aka. Friedrich Wertheimer), a German Marxist psychoanalyst emigree, in the article *"Calling All Couriers"*,[114] published in the then Soviet-controlled and highly influential *New Republic* magazine. Wertham's article is a malicious "book review" of Reich's *Mass Psychology of Fascism*, objecting to his penetrating criticisms

In Defense of Wilhelm Reich

of Soviet communism, declaring *"Of Reich's Marxism little remains but a hollow shell"*. Wertham is a known crusader for press censorship of children's comic books and racy novels, arguing for authoritarian child-raising. *New Republic* is then edited by Soviet sympathizer Henry Wallace and published by Michael Straight, a cloaked American member of the notorious *Cambridge* spy ring, which took atomic and other secrets to the Soviet Union. Reich is also approached for an interview by Mildred Edie Brady (he refuses), a cloaked communist and Soviet mole, who will soon thereafter slander Reich in the same *New Republic*.

1947: Reich starts a new journal, *Annals of the Orgone Institute,*[115] with major new articles: *"Work Democracy in Action"* and *"The Expulsion of Wilhelm Reich from the International Psychoanalytic Association".*[6] His research continues. An *orgone motor force* is discovered, and investigations suggest an *anti-nuclear radiation effect* in the orgone energy. Very high count-rates are observed in nuclear radiation detectors when they are charged within an orgone accumulator. Reich begins construction of the observatory building on his new property in Rangeley, Maine, which he calls *Orgonon*.

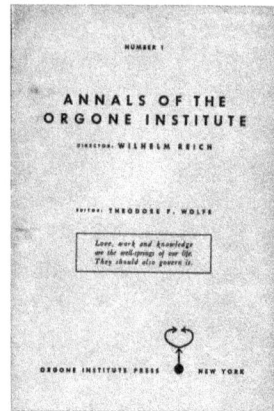

1947 ATTACKS: A massive new smear campaign against Reich begins, triggered by the above-mentioned Communist writer Mildred Brady. She writes two articles attacking and slandering Reich: *"The Strange Case of Wilhelm Reich"* in the *New Republic,* and *"The New Cult of Sex and Anarchy"* in *Harpers.*[116] Brady's articles (especially in *N.R.*) heap scorn and sarcasm upon Reich and his discoveries on human sexuality and cosmic energy. She fabricates sexual slanders, claiming Reich prescribes the orgone accumulator to "derive orgastic potency". She repeatedly sarcastically denigrates orgastic concepts, revealing her anxious contempt for sexuality. Reich's findings on character analysis, emotional armoring, and particularly on the consequences of undischarged sexual tension and the function of the orgasm are all denigrated by Brady. She defends Stalin's Russia against Reich's observation that it is a sex-negative authoritarian culture, and he is additionally slandered as leading a purported but nonexistent "orgone cult". She ends with a call for government action against Reich, and thereafter works behind the scenes towards that goal. Mildred and Robert Brady are known confidants and friends of the Soviet spy ring led by Nathan and Helen Silvermaster, as well as with Soviet spy Norman Bursler, who once shared an apartment with Robert Brady. Around the same time when the Brady articles are appearing in the press, the FBI has the Bradys, Bursler and the Silvermasters under surveillance. They gather for dinners and parties at the home of the Silvermasters, who also welcome various East Bloc diplomats into their home. Nathan Silvermaster will later receive a special Soviet Award for his spying activities.[117]

The cloaked Stalinist psychiatrist and Reich-hater Karl Menninger, follows up on his 1942 slanders against Reich by reprinting the Brady article verbatim in his *Menninger Clinic Bulletin*,[116] giving it a false "medical-scientific" credibility. Menninger also has many articles published in *The Nation* and *New Republic* magazines,[118] the two most visible socialist-communist publications in America at the time.

Brady and Menninger, along with other psychoanalysts, psychiatrists and doctors who hate Reich and his clinical findings, use their influence to persuade the *US Food and Drug Administration* (FDA) to start what quickly degenerates into a *Get Reich* sham investigation.[1,51,119] New lies and slanders appear in major magazines and newspapers, repeating or drawing upon the Brady article (i.e., *"The Strange Case of Wilhelm Reich"* in *Everybody's Digest*, *"The New Coast of Bohemia"* in *Saturday Review*, and *"Beware of the Medical Frauds"* in *Saturday Evening Post*).[63] Multiple other copy-cat articles and book-chapters appear thereafter, including within the influential and left-leaning *Time* magazine.[63]

Reich and his associates write rebuttal articles and letters-to-the-editor against this onslaught, but few are printed. The chorus of slander and smears continues to grow.[63]

Copies of the Brady articles are sent to the FDA's East Coast offices, which undertake a preliminary investigation, interrogating Brady and others. At first, Reich cooperates with the FDA investigation, asking his associates to do the same. However, the FDA is snooping for a "sex-cult" based upon the Brady smears. When Reich learns of this, he instructs everyone to cease cooperating and

Orgonon, Reich's Home and Laboratory

In Rangeley, Maine. *Upper*: The Observatory with Reich's laboratory, archive, library, study and observation platforms. *Lower*: The Student's Laboratory. Reich moved his facilities to this remote location, given its clean air and excellent conditions for orgonomic research. He had plans to build a clinic or hospital, as a center for medical therapy and training. Instead, it was shut down and Reich was destroyed by a power-drunk US Food and Drug Administration. Today it is open to the public as the *Wilhelm Reich Museum*, offering a bookstore and other facilities. http://www.wilhelmreichtrust.org

instead to refer the FDA to him. FDA Chief Inspector Charles A. Wood secretly romances Clista Templeton, the daughter of one of Reich's assistants, recruiting her to become a spy for the FDA; they are married a short time later.[1,119]

By December, the FDA has managed to interrogate one of Reich's assistants, the mathematician William Washington, pressing him to reveal details of Reich's experimental physics work. The FDA agents, presumably only mandated to investigate possible violations of food and drug laws, circulate internal memos excitedly asking for exact detailed information on Reich's *orgone motor* and Geiger-counter experiments. Reich becomes aware the FDA is influenced by communists (ie, Mildred Brady and others) and out to steal the secrets of his orgone motor and later anti-nuclear (oranur) discoveries. Few believe him.[52,119] However, internal FDA files obtained years later prove Reich is correct:

"It will be quite helpful, I believe if we can get from [William] Washington the exact details of the experiment...when he stated that the Gerger (sic) Counter registered increased cosmic ray particles and orgone particles when attached to the Orgone Accumulator... If we can get in detail how this experiment was performed, including all attachments, physical setup, positions, and other details, including the model of the Gerger Counter used, with also the details ... the sizes of accumulators used...the number of times the experiment was repeated, if he has figures of comparative results, we should try to get these figures." Letter from W. Wharton, Chief of FDA Eastern District, to K. Harris, Chief of FDA Boston Station, 10 Dec. 1947.

"I have noted your report of what William Washington says concerning cosmic ray counts and orgone ray counts. When you next see him, try to get him to explain the difference between these two alleged rays, and how much they are shown differently on the Geiger Counter. Ascertain as much as you can about the experimental work that was done with the Geiger Counter in Rangeley, Maine when Washington was there." Letter from W. Wharton, Chief, FDA Eastern District, to Inspector C. Wood, FDA Boston Station, 12 Jan. 1948.

In Defense of Wilhelm Reich

Reich consults with a well-known New York attorney, Arthur Garfield Hays, to launch a lawsuit against the various writers and government agents spreading defamatory slander against him. Reich writes to Hays: *"I wish to stress the fact that the slandering article by Miss Brady was the beginning of a chain reaction set into motion, beyond any doubt, by communist quarters. The husband of Miss Brady is a communist."*[120] Unknown to Reich and as detailed previously, Hays is a Soviet sympathizer intimate with many communist-front organizations and individuals, including the FDA advisory group *Consumer's Union*, and with the Bradys. Reich's declaration against Brady and other communists apparently unsettles Hays, who dissuades Reich from taking legal actions against his slanderers, unethically failing to disclose that he knows and has intimately worked with some of them.[117]

1948: Reich continues with his research, attracting new students and associates. He publishes *The Cancer Biopathy*,[121] which presents his findings on cancer as a systemic biopathic disorder. Reich argues how the appearance of tumors is only the end-phase of the disease. He presents evidence on how the process of emotional resignation, sexual stasis, respiratory blocking, energetic depletion, and finally the bionous disintegration of the patient's own tissues, leads to the appearance of bionously reorganized amorphous cancer cells and tumors. His treatments of cancer patients employ methods to increase respiration and emotional expression, addressing sexual difficulties, and use of the orgone accumulator to restore charge to depleted tissues. He notes how these methods worked best to prevent the energy-loss typical of the cancer process, and how the orgone accumulator can disintegrate tumors and extend life. He also notes how this can also result in death from the breakdown detritus of disintegrating tumors, which clog the liver and kidneys.§ The accumulator is not promoted as a "cancer cure". Reich transfers all rights for medical use of the orgone accumulator to the nonprofit *Orgone Institute Research*

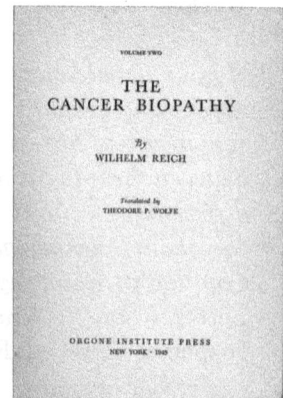

VOLUME TWO

THE
CANCER BIOPATHY

By
WILHELM REICH

Translated by
THEODORE P. WOLFE

ORGONE INSTITUTE PRESS
NEW YORK · 1948

---

§ Today, new methods of detoxification can beneficially work with the orgone accumulator to improve outcomes.[96]

Laboratory. The *American Association for Medical Orgonomy* is formed by Reich and several of his physician associates.

1948 ATTACKS: Reich publishes *Listen, Little Man!,*[122] a rebuttal of his many irrational critics. Reich's associate Theodore Wolfe also writes a book exposing the malicious lies in the press: *Emotional Plague Versus Orgone Biophysics: The 1947 Campaign.*[123] These have no appreciable effect to stop the public attacks. Internal FDA memos indicate they cannot find any dissatisfied users of the orgone accumulator. They try to get the US Post Office to halt deliveries of Reich's books, especially *The Sexual Revolution,*[111] under anti-pornography laws, because the book openly discusses masturbation. Reich is also attacked in the book *Communism and the Moral Breakdown of America* by Albert Crombie, an angry Christian moralist who was inspired by the Brady slander articles. Crombie sends letters to the *Immigration and Naturalization Service* (INS) claiming Reich is a communist pervert who should be deported. The INS begins an investigation of Reich, soon confusing him with a real communist radical in New York City named *William* Reich. Only in 1956 is this confusion finally cleared up and the investigation closed.[100,119,124]

One of Reich's physician associates, Walter Hoppe, is detained at Ellis Island when travelling from Israel to attend an Orgonomic Conference Reich has organized. Reich officially protests and Hoppe is released.[1,119]

1949: Reich retitles his journal as the *Orgone Energy Bulletin,*[125] to reflect an increased focus upon his biophysical life-energy research. In spite of the many attacks, it is a very productive period for Reich. New experimental papers are published on the physics of orgone energy functions: *"Orgonomic Functionalism, Part I: Ether, God & Devil", "Orgonotic Light Functions 1: Searchlight Phenomena in*

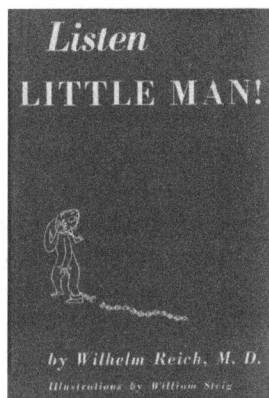

Health Benefits of the Orgone Accumulator

Reich noted the orgone radiation, as intensified within the orgone accumulator, would stimulate the parasympathetic nervous system. Overall, this had a counter-effect upon any tendencies towards chronic contraction or anxiety. It helped the organism to expand, brought a glowing warmth to the insides, and stimulated other parasympathetic reactions. These effects were tested out in two separate double-blind and controlled studies undertaken at European universities, one by Müschenich-Gebauer in 1986 at the Univ. of Marburg[126] and another by Hebenstreit in 1989 the Univ. of Vienna.[127] Both studies confirmed Reich's observations with moderate to high levels of statistical significance. During and after Reich's time, numerous clinical case-studies were published showing clear benefits from the accumulator, including *symptom reductions* (but no "cure") for a variety of conditions and disorders, including cancer.

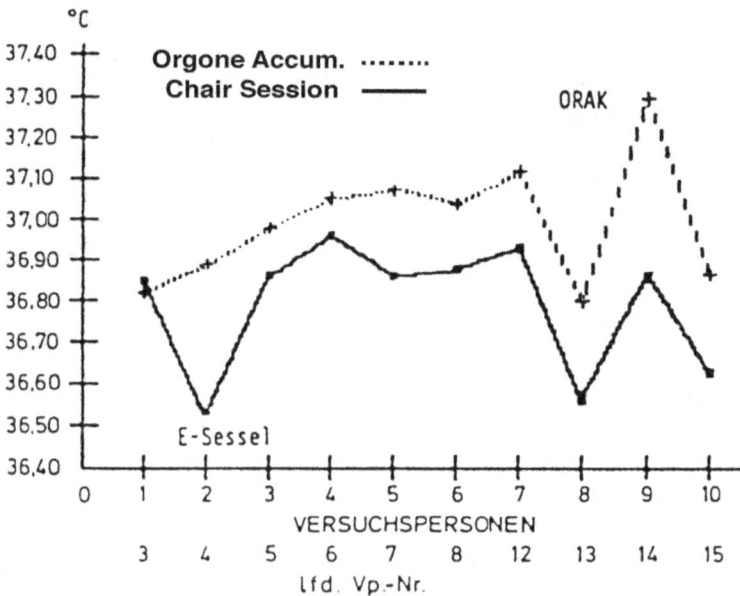

Average body-core temperature of test subjects during session of sitting inside the orgone accumulator (top line) versus prior sessions of sitting inside a dummy box.[126]

the Orgone Energy Envelope of the Earth", "A Motor Force in Orgone Energy", "Orgonotic Light Functions 2: An X-Ray Photograph of the Excited Orgone Energy Field of the Palms", "Orgonotic Light Functions 3: Further Characteristics of Vacor Lumination", "Cosmic Orgone Energy and 'Ether'". [6]

1949 ATTACKS: The FDA's "investigation" of Reich is fully underway by now, but mostly concealed from view. More attacks also appear against Reich within influential allopathic medicine and leftist publications: *"Cancer and the Need for Facts: Orgone Accumulator" (Journal of the American Medical Association)*;[128] *"Cancer 'Cures' Beware" (Consumers' Report).*[129] Reich's physician associates Chester Raphael and Elsworth Baker are also attacked and fired from their positions at Marlboro State Hospital.[130] Reich and his associates write clarifying rebuttal articles and letters, to little affect. Reich's assistant William Washington, having previously been solicited by the FDA to provide details on Reich's orgone motor, now disappears with one of the motors and other components.

1950: Reich continues to work, and publishes many new experimental papers: *"Orgonometric Equations I: General Form", "Meteorological Functions in Orgone-Charged Vacuum Tubes", "Oranur Project: The Orgonomic Anti-Nuclear Radiation Project", "Orgonomic and Chemical Cancer Research: A Brief Comparison".*[6] He also does not neglect the social issues, in other papers reflecting such work: *"About Genital Self-Satisfaction in Children", "Children of the Future, I: Report on the Orgonomic Infant Research Center."*[6] Reich's medical associates are also treating people with the orgone energy accumulator for various diseases, and many of their papers are published in the *Orgone Energy Bulletin.*[6,124] Reich observes a powerful healing of burns with the accumulator.

Reich's findings by this point are documenting the existence of an energy continuum filling the atmosphere and also present in high-vacuum, suggesting it also exists in open interplanetary space. He has developed a small motor to run from this energy, and has a new theory which postulates the ionization within classical radiation-detection equipment is an orgone energy effect. He is also documenting numerous orgone energy functions at work in atmospheric processes and weather. All of this is breakthrough research.

In Defense of Wilhelm Reich

1950 ATTACKS: More slander and smears attacking Reich appear in popular books and articles. These include the first of several writings by the Stalinist pop-writer and false "skeptic", Martin Gardner, *The Hermit Scientist* in *Antioch Review*.[131]

1951: Reich publishes the books *Ether, God and Devil*[132] and *Cosmic Superimposition*.[133] In these he notes the antithesis between mystical religion and mechanistic science, each of which stands in opposition to the other, but with a *common functioning principle* as discovered in the orgone energy. The orgone is an omnipresent, pulsating *life-energy* that fills all space. Reich also discusses the role of *superimposing orgone energy streams* in the creation process, as with the formation of galaxies, hurricanes, life and raw matter from the cosmic background.

Reich publishes *The Orgone Energy Accumulator, Its Scientific and Medical Use,*[134] summarizing the principles of orgone energy research and discoveries for the layperson. His work on the anti-nuclear radiation effects of orgone energy is also published in *The Oranur Experiment.*[135]

In this experiment, one milligram of radium is brought inside a strong orgone accumulator, creating a powerful energetic reaction in the laboratory. People working in the lab fall sick, and the building is enveloped in a dull blue glow. An anomalous fever-like reaction spreads from the laboratory over the adjacent landscape, even after the radium is removed to a far distance. The spreading field effects also trigger local weather changes, while laboratory mice and trees near Reich's lab begin to die. Laboratory workers recover only by leaving the area, and the labora-

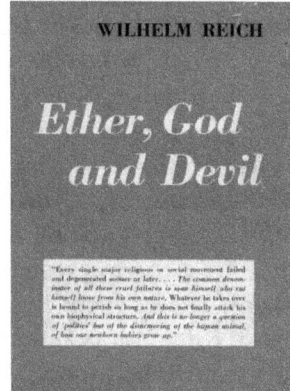

tory itself becomes uninhabitable. Reich thereby becomes one of the first scientists to detail extreme toxic effects from low-level atomic radiation. The properties of the orgone-influenced radium are also changed. It is later ascertained that exposure to this *orgone anti-nuclear effect (oranur)* triggers an inflammation, but also a healing of old wounds and injuries, indicating a life-positive aspect similar to but more powerful than what is today termed *radiation hormesis*. Reich sends copies of his *Oranur Experiment*[135] report to various government officials, but he is ignored.

Reich and his associates publish more papers on *orgonometric equations*, orgone-charged high-vacuum *vacor* experiments, weather effects of orgone energy, as well as others on experimental orgone accumulator therapy of people with various diseases.[6]

Reich and Ilse Ollendorff separate and later divorce.

1951 ATTACKS: Through an intermediary, Reich approaches the influential Eleanor Roosevelt, widow of the late President Roosevelt, in efforts to gain scientific support for his important new findings on oranur. She contacts Robert Oppenheimer for an evaluation, and he calls Reich's work "a hoax", which sabotages the effort. Oppenheimer's top secret security clearances are revoked a few years later, partly for his connections to the Stalinist-run Consumers' Union, and his friendly relations with Mildred Brady.[90,91]

More smears against Reich are published, parroting the original Brady smears of 1947 (i.e., *"Cancer Quacks"* in *Collier's*; *"Are Psycho-analysts Crazy?"* in *Cosmopolitan).*[63] FDA agents continue working toward Reich's destruction with their biased "investigation".

1952: Reich's productive work continues, with new articles including many by his medical and scientific associates in the *Orgone Energy Bulletin*, such as: *"A Space-Energy Continuum"*, *"Genital Anxiety in Nursing Mothers"*, *"Observations of Orgone Energy Lumination"*, *"Observations on a Case of Coronary Occlusion"*, *"From Libido Theory to Orgonomy"*, *"On the Medical Use of Orgone Energy".*[6]

Oranur reactions continue to make Reich's laboratory uninhabitable. The high levels of energy charge and excitation are supplanted by a stagnant quality, which Reich calls *DOR*, short for *deadly orgone radiation*. In efforts to drain the DOR phenomenon off into water, the *cloudbuster* is developed – developed from the observed affinity of the orgone energy to water. The cloudbuster

The ORgone Anti-NUclear Radiation Effect: ORANUR and VACOR (Orgone Charged Vacuum)

Reich's experiments indicated the orgone was a cosmic energy with lifelike properties, filling all space but previously undetected by physics or biology. He tested the orgone radiation with various kinds of standard energy detectors, finding that most would not register anything. However, Geiger-Müller tubes would react if allowed to soak in the orgone accumulator for an extended period. This led to experiments with orgone-charged high vacuum tubes at around 0.5 micron pressure, which showed anomalous reactions not anticipated by conventional physics. These experiments indicated there is no "empty space". Orgone research, like that into "dark matter", "neutrino sea", "cosmic ether" or "cosmic plasma", proved the background of open cosmic space is *energy-rich*. In the historic *Oranur Experiment*,[135] Reich observed changes in the decay-rates of radioactive isotopes.

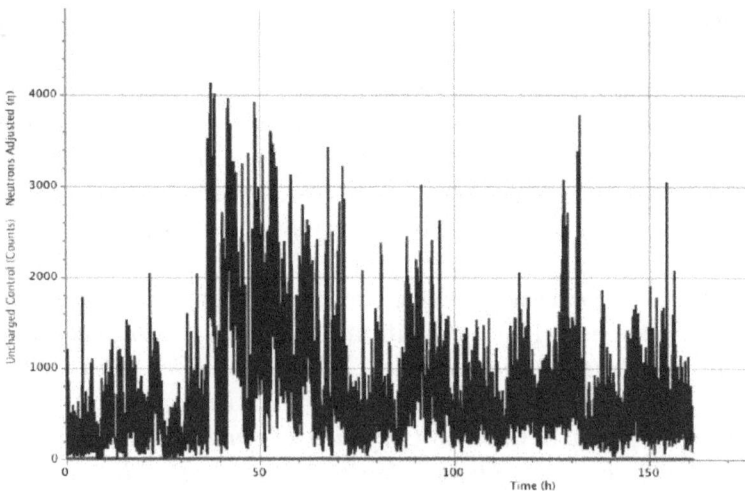

Very high count rates from a special thick-walled GM detector tube, soaked for a year inside a strong orgone accumulator at the author's laboratory. No radiation sources were used, background only. Normally this yields about 5 cpm, but with orgone charging it yielded up to 4000 cpm.[136]

proves effective to end the crisis. However, the new cloudbuster device can also intensify cloud growth to bring rains. Reich concludes that *drought* has a deeper-lying energetic component based upon chronic DOR, which suffocates life, dries out the atmosphere and works towards the development of deserts. He identifies a blackish precipitate from an acidic, dor-infused atmosphere which kills trees and blackens rock; he is the first scientist to identify critical components of what later is called *acid rain* or *forest death*.

Reich makes an audio recording expressing his feelings of isolation and aloneness, that few of his medical associates have the capacity to understand and follow him into this new scientific territory.[137]

Chester Raphael and Helen MacDonald publish a summary of Reich's seminar: *Orgonomic Diagnosis of Cancer Biopathy*.[138] Kurt Eisler of the *Freud Archives* conducts an interview with Reich, published years after his death as *Reich Speaks of Freud*.[139]

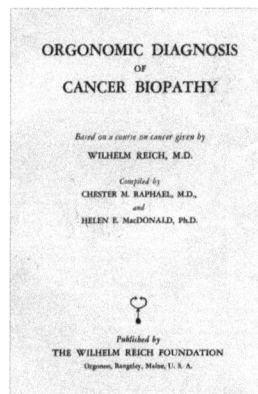

ORGONOMIC DIAGNOSIS
OF
CANCER BIOPATHY

Based on a course on cancer given by
WILHELM REICH, M.D.

Compiled by
CHESTER M. RAPHAEL, M.D.,
and
HELEN F. MacDONALD, Ph.D.

Published by
THE WILHELM REICH FOUNDATION
Orgonon, Rangeley, Maine, U. S. A.

1952 ATTACKS: The smear attacks which began in 1947 continue in the yellow press and tabloids of the day, all repeating and amplifying upon the original Brady slanders (i.e., *"The Strange Case of the Orgone Box"* in *Man's Magazine*[63]). The Marxist-Humanist Martin Gardner also joins in the mudslinging, with distortions and slander in a chapter on *"Orgonomy"* included within his book *In The Name of Science*,[140] which is later retitled *Fads and Fallacies In the Name of Science*. This book is reviewed in academic journals and placed in libraries, so his outreach is far and wide. A few years later, in 1956, Gardner gains a foothold and column within *Scientific American* magazine, where his smears and lies on unorthodox research of all kinds are spread internationally. He constantly repeats the Brady smears against Reich, and cooks up new ones. As is typical of these attacks, none of the original experimental works nor the verification studies by others are mentioned, in deliberate *lies of omission*.

The US State Department and INS continue to secretly investigate Reich's activities in Norway, querying if he had been a member of the *Norwegian Communist Party*. An on-station CIA agent in Norway investigates, and finds the accusation against Reich is

In Defense of Wilhelm Reich

Wilhelm Reich, 1952, USA

In the midst of major press attacks slandering his name and work, under pressure from a ruthless government investigation aiming to utterly destroy him, Reich nevertheless made numerous new scientific discoveries of a breakthrough nature. He was fully aware of the dangerousness of his situation, but also of his responsibilities to defend his new research findings. His enemies were legion, his friends few. No scientific or "academic freedom" organizations came to his aid. Even today they try to hide the fact that they either did nothing, or joined in with the lynch mob. Reich declared he would not go quietly like Socrates, and so stood on moral and Constitutional principles, expecting the US Courts to do likewise. They did not, and in the end Reich died in prison, his books burned by *order* of those same courts.

false.[100,124] That part of the investigation is dropped. Meanwhile, the FBI receives letters from people in Rangeley, Maine, Reich's hometown, expressing suspicions about "foreigners" visiting Reich's institute, based upon what they read in various newspaper and magazine smear articles. A local Rangeley druggist, influenced by the smears, organizes a demonstration of teenagers outside Reich's gate, shouting *"Orgy, Orgy, Commie, Commie"*. Reich confronts them and they leave.[1,119]

1953: Reich publishes new books: *People in Trouble*[141] and *Murder of Christ.*[142] His *Orgone Energy Bulletin* continues to print many papers and reports, as well as rebuttals to smears. Reich also publishes the correspondence between himself and Albert Einstein, in *The Einstein Affair.*[98] He holds a seminar on *DOR, Drought and Desert,* and demonstrates the *medical dorbuster,* a new therapeutic device for removing energetic stagnation from the body. He begins a discussion on the *Emotional Desert,* and publishes *"The Blackening Rocks"*,[143] the first scientific report on what is years later described by classical science in terms of air pollution, acid rains and forest-death. He makes a successful demonstration of the cloudbuster to local blueberry growers, who fear crop failure due to drought. Reich is also studying the UFO phenomenon, having seen several near his rural laboratory in Maine.

1953 ATTACKS: Reich is interviewed by the INS, as part of their continuing investigations into his past communist associations.[1,119] They ignore his evidence of anticommunism since c.1933, and also the conspiracy of public slander directed against him by cloaked communist American writers. But they do ask if he will inform on other ex-communists; Reich refuses.

Reich's Discovery of the Atmospheric and Cosmic Orgone Energy[143,144,145,146,150,151,152]

- First discovered as a radiation from sand bions, isolated within an insulated metal *Faraday*-type enclosure, which was later developed into the *orgone energy accumulator*.
- Experiments proved the energy was a radiant excitable and mass-free energy continuum, filling all space, similar to the cosmic ether, except being motile, pulsatile and sensible.
- Orgone energy has a negative entropy and will concentrate to higher levels where possible. The orgone accumulator thereby yields thermal, electrostatic and humidity anomalies. It can charge up objects to yield other anomalies, as well as having healing effects upon burns or wounds, with symptomatic benefits to various biopathic, degenerative health conditions.
- The energy flows according to spiral-wave characteristics, the *Kreiselwellen* as Reich observed, which has a mathematical regularity he described in new *orgonometric equations*.
- Reich's theory of *Cosmic Superimposition* postulated spiraling and merging orgone energy streams as the basis of both hurricane and galaxy formation. Superimposition also functioned at the microscopic and cellular levels, in the creation of matter and underlaying sexual attraction and procreation.
- The affinity of orgone energy towards water indicated its regulatory function within clouds, weather dynamics and the atmosphere, a fact which was later proven by experiments with the cloudbuster. Reich provided the first scientific discussions on acid rain, forest death, and hazy drought/desert atmospheres.
- Reich's ideas are agreeable with much of the older theory of cosmic ether, and both predated and anticipated such physical concepts as the *neutrino sea, cosmic plasmas, interstellar medium,* and *dark matter* – many of which are identified by similar blue-glowing and ubiquitous characteristics. He is rarely cited for his scientific priority, however.
- The moving and streaming *cosmic orgone* was described by Reich as comparable to the *cosmic ether,* as a prime mover in astrophysics in the Galilean sense. Reich's orgone theory is agreeable with the *empirical* foundations of modern astronomy, but challenge theoretical concepts such as Einsteinian relativity, big-bang creationism, and quantum metaphysics.

1954: Reich undertakes new experimental work with the cloudbuster device, ending a major drought in the NE USA. He again changes the name of his research journal to *Cosmic Orgone Engineering (CORE),*[144] reflecting his new focus upon the atmospheric functions of cosmic energy and the new cloudbusting methods for ending drought. The first issue of his *CORE* journal carries the extensive report *OROP Desert.*[145] His papers focus upon the role of energetic stagnation (DOR) in desert and drought formation.

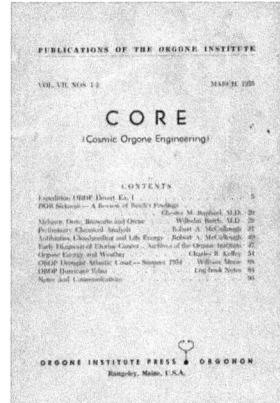

Reich observes UFOs near his Rangeley laboratory, and aims the cloudbuster at them, after which they dim-out and vanish, or silently race away. Reich sends information on the orgone motor and his UFO sightings to the US Air Force, speculating they are spacecraft propelled by orgone energy. His assistant and son-in-law Robert Moise (husband of Eva Reich) meets with the *Air Technical Intelligence Command* (ATIC) in Dayton Ohio, discussing Reich's findings on oranur, the orgone motor, his orgonometric equations and his theories about UFOs, gravitation and negative-gravity. Reich's publications *OROP Desert*[145] and later *Contact With Space*[146] include reports on these observations and meetings.

In October, Reich undertakes a major desert-greening experiment around the Tucson Arizona region with his new cloudbuster method. The experiment, which ends only in July of 1955, is a success, bringing moisture, rains and increases in natural grass over the region. More UFOs are sighted, however, and Reich believes the cloudbuster is somehow attracting them, possibly affecting the energetics of their propulsion.[146]

A new journal, *Orgonomic Functionalism*, is started in the UK by Paul and Jean Ritter, running for the next 10 years.[6] This gives some small voice primarily to the British followers of Reich's work. Reich is uncomfortable with this, however, as it is published without his oversight, and he is rightly concerned about distortion of his ideas.

1954 ATTACKS: The FDA continues snooping and spying on Reich, and now launches a legal prosecution. The FDA goes to the

In Defense of Wilhelm Reich

Federal Courts petitioning for a *Complaint for Injunction* against Reich,[1,119] to stop his work and activities, and halt movement of his books and orgone accumulators across state lines, via the Commerce Clause. Peter Mills, one of Reich's former personal attorneys, now unethically appears as the lead prosecutor in the FDA case. Reich writes a formal *Response*[147] to Federal Judge John D. Clifford, detailing the lies and dirty tricks of the FDA, the falsity of the claims against himself and evidence in favor of his research findings. In this *Response,* he also refuses to *"...appear in court as the 'defendant'... in matters of natural science."*[147] Judge Clifford lazily, arrogantly and *criminally* ignores Reich's *Response*, acting as if *no response* had been given whatsoever. Having neatly discarded Reich's *Response*, and fully embracing the FDA's case, which is based upon the Brady lies, Judge Clifford compounds his arrogant breach of justice by refusing to investigate further and declaring, without scientific expertise or justification, that *"orgone energy does not exist"*. By default, the orgone energy accumulator is dismissed as *"mislabeled merchandise"*. Reich's books are then redefined, by Clifford's ex-cathedra judicial decree, as *"advertising literature"* and ordered to be destroyed, no longer being "real books" protected by the *US Constitution's First Amendment*. The FDA is thereby handed a technical "victory" and is given 100% of what they ask for, including *the court-declared anti-Constitutional banning and burning of Reich's books*, the so-called "advertising literature". A command is also given that Reich cease speaking and writing about his research findings.[1,119] New smear articles appear afterwards in the mainstream press, gloating over Reich's tragedy (i.e.: *"The Strange Case of Dr. Wilhelm Reich"*, in the *New York Post* [63]).

Various "mental health professionals" who had helped the FDA in their prosecution against Reich also exchange congratulatory letters, celebrating Reich's conviction:

> *"We are most appreciative of your letter of March 25 relative to action taken in connection with Dr. Wilhelm Reich and his group. The American Psychoanalytic Association wishes to commend the Food and Drug Administration for their effective action in this situation."* (Letter from Richard Frank, Secretary of the American Psychoanalytic Association, to the FDA.[119, p.152])

Field Trials with a *Reich Cloudbuster*

While the Reich cloudbuster may superficially appear to be an "impossible device", it has been subjected to systematic experiments and field tests, with confirming positive results. The author undertook a 2-year study at the *University of Kansas* in the late 1970s, showing it could dissipate cumulus clouds as compared to control clouds, and also increase rainfall across the State of Kansas.[150] Later systematic work was undertaken in Arizona, showing a *rainfall-doubling effect*. Major drought-breaking field experiments were undertaken in Israel, Namibia and Eritrea, Africa, by the author and on invitations from their governments.[151, 152] Droughts were ended quickly, well beyond the weather forecasts, and landscapes greened. At every turn, however, the mechanistic meteorologists were against the various projects, attempting to cut their funding, often spreading slanders, and discarding their objectivity. Today, all this work is nearly censored out of existence and slandered with falsehoods, outright lies, and lies of omission, from the notoriously unreliable *Wikipedia* to the "scholarly" *Weatherwise* magazine.

http://www.orgonelab.org/ResearchSummary2.htm

In Defense of Wilhelm Reich

"We are delighted to hear of the successful prosecution of your action against the Wilhelm Reich Foundation, and I know that I speak for the profession at large in expressing our deep appreciation of the good work of the Food and Drug Administration." (Letter from Daniel Blaine, Medical Director of the American Psychiatric Assocation, to the FDA.[119, p.152])

"The acting Medical Director of the Federal Food and Drug Administration has expressed his Agency's appreciation for APA's help in the successful development of its case." (Reference to a letter from the FDA, published in the *American Psychiatric Association Newsletter* of 15 April 1954.)

Reich thereafter details the larger scope of the conspiracy, exposing his enemies in psychoanalysis, psychiatry, and from within leftist-Communist circles, in a new documentary publication: *Conspiracy: An Emotional Chain Reaction*.[148]

1955: Through his *CORE* journal, Reich publishes more evidence supporting his discoveries: *"Expedition OROP Desert, Ea, 1: DOR Clouds Over the USA", "The Medical DOR-Buster", "OROP Hurricane Edna", "Early Diagnosis of Cancer of the Uterus", "Priority of Wilhelm Reich's Cancer Findings"*.[6,144] Reich continues his Tucson research into desert-greening, which ends in April. He begins a romance with Aurora Karrer, a biologist from the *National Institutes of Health* (NIH). Reich's medical associates begin publishing a new journal, *Orgonomic Medicine*,[149] which includes case histories of their findings on orgone therapy, including medical use of the orgone energy accumulator.

1955 ATTACKS: In January, while Reich is still working in Tucson Arizona and without Reich's knowledge or permission, his associate

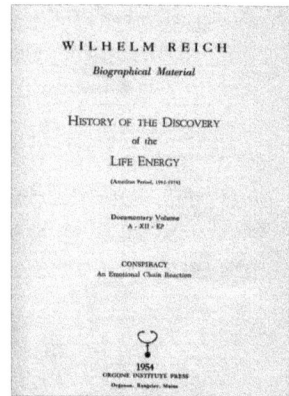

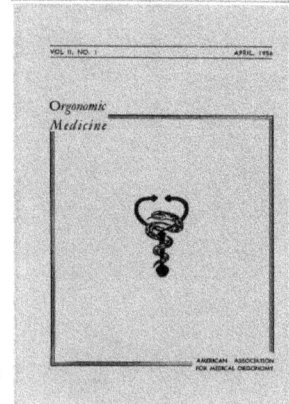

Michael Silvert takes a truckload of books and orgone accumulators from storage in Maine, to storage in New York. This is a technical violation of the court-imposed injunction against crossing state lines with "mislabeled merchandise". The FDA learns of the "unpermitted transportation" of books and accumulators, and investigates. FDA agents accompanied by State Marshals visit Reich's base of operations in Tucson, interrogating him. In July, Reich and Silvert are charged with "contempt of court" for technical violation of the FDA injunction.[1,119]

Reich publishes two documents addressing his critics and the new challenges by the US Courts and FDA: *Response To Ignorance,*[153] and *The Red Thread of a Conspiracy.*[154]

1956: Reich publishes more on the conspiracy against him: *Atoms for Peace Versus the HIG (Hoodlums In Government).*[155] He submits to the Judge handling his contempt case, that he is the anonymous unnamed scientist mentioned in *Report on Flying Saucers,*[156] authored by Major Edward J. Ruppelt, head of the US Air Force Technical Command's *Project Bluebook*. Reich also prepares two volumes, *Suppressed Documentary Evidence*[157] and *Suppressed and Top Secret Evidence,*[158] which are sent to various officials and to the Judges. *It is all ignored,* as is the issue of book burning.

1956 ATTACKS:[1,119] Reich's contempt-of-court trial begins in Portland Maine under Judge George C. Sweeney in early May. As noted above, his former personal attorney, Peter Mills, unethically appears as the prosecuting attorney. Reich is now consumed with the legal fight, as he is physically arrested and dragged into court, forced to post bail and to defend himself for violation of the fraudulently-obtained FDA injunction. Judge Sweeney refuses to take any testimony on the question of the orgone energy or on deceit by the FDA, which by the legal procedures was "declared" to have been "resolved" in the original injunction (where Judge Clifford threw Reich's *Response* into the trash can). Reich is repeatedly silenced in the court through procedural truth-killing, and is convicted. He is sentenced to two years in prison, with Silvert getting one year plus one day, a guarantee they will both lose their medical licenses. The *Wilhelm Reich Foundation* is fined $10,000.

Reich petitions the Appellate Court for reversal of the conviction, which is denied. He then appeals to the US Supreme Court, moving

In Defense of Wilhelm Reich

to Washington DC in efforts to carry forward with his appeals.

Reich's remaining associates try to defend him in the courts with a *Writ of Certiorari,* saying their work will be disrupted if the orgone accumulator is forbidden.[159] The court rules, however, that it has *no interest in what others do with the orgone accumulator; they are only interested in what Wilhelm Reich is doing* (!). This affirms the whole procedure was set up to specifically *Get Reich,* and *Destroy Reich*, and is thereby indicative of foul play and political deceit transpiring "behind the curtain".

CONTACT WITH SPACE

ORANUR
SECOND REPORT, 1951-1956

OROP DESERT Ea
1954-1955

by
WILHELM REICH

CORE PILOT PRESS
New York, N. Y.
1957

FDA agents now seize truckloads of Reich's books and research journals, eventually carting off six tons of them to incinerators in New York City. FDA agents invade Reich's Orgonon facility and put orgone accumulators to the axe. More smear articles appear, in both mainstream and tabloid publications (i.e., *"Sex Cult With Class"* in *Sir!, "Learn to Read the Faces of Foxes, Dried-Out Lemons, Haters of Love"* in *Washington Daily News, "Quieting the Quacks"* in *The Daily Oklahoman).*[63]

1957: Reich's last book *Contact With Space*[146] is published. In it, he develops his theory of cosmic influences upon atmospheric weather, and the origins of drought and desert in energetic stagnation, detailing the results of his cloudbuster experiments in Arizona. He also speculates on ways to green large deserts and describes numerous observations of silent UFOs, the flight-tracks of which are photographed. He speculates they are powered by the cosmic orgone energy. He observes a noxious energetic "exhaust" when they are in the vicinity, indicative of DOR, and describes how several of them visibly dim in luminosity when the cloudbuster is aimed at them. The book also gives new orgonometric equations and information about *gravimetric orgonomy*, the *KRx Law of Pendulums*, and other things which may ultimately power new spacecraft and lead a future rational humanity out to the stars.

1957 ATTACKS:[1,119] Reich loses his appeal to the US Supreme Court, which declines to review the unethical, anti-Constitutional decisions and actions of the lower courts. The court-case wrap-

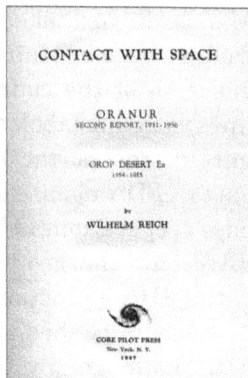

pers[160] later indicate the Supreme Court Judges are unconcerned about the FDA lying to the lower courts, and have no objections to the burning of Reich's books. They express an arrogant attitude towards Reich for not following proper court etiquette (like a King on his Throne, contemptuously dismissing an unwashed peasant bound in chains, because he didn't address the King by his proper title, and doesn't bow sufficiently low). The Court also specifically noted the Brady-Gardner slanders as "facts" underlying the case, in full agreement with the FDA's lying (*"It is asserted by the Government that these devices were being falsely held out to the public at large by the defendants as at least beneficial in the treatment of a great number of human ills ranging from cancer to the common cold"*), and referenced the prior rulings of the Vienna-born, Communist and pro-Soviet Judge Felix Frankfurter,[94,160] who previously had written into law a most astonishing de-facto *"Doctrine of Judicial Infallibility"*[160] (my words). This amazing legal proclamation asserted that court rulings in contempt proceedings *can not be questioned*, even on the basis of prosecutorial lying and suppression of facts as Reich had documented, nor as regarding judicial overreach in the original ruling to require, for example, *anti-Constitutional actions such as the burning of books.*

Frankfurter is a sitting Justice on the US Supreme Court at the time of Reich's appeal, having been appointed by Roosevelt in 1938. He heavily sways the Supreme Court to accept the lower court's ruling against Reich, which cited his own preceding case ruling about "infallible judges". A longtime advisor to President Roosevelt (along with Soviet mole Henry Wallace and the pro-Soviet Arthur Hays), Frankfurter is a primary author of New Deal legislation and advocate of

Reich's Arrest, March 1957

85

In Defense of Wilhelm Reich

judicial activism with the central theme of empowering Big Federal Government over the rights of the individual and States. He also argued for *repeal of the Due Process clauses in the US Constitution's 5th and 14th Amendments*, guaranteeing all citizens the rights to *"life, liberty and property"*.[94] This was so, even as he wrapped himself in the aura of being a "civil liberties advocate". The same was true with Reich's former attorney Arthur Hays, who worked with Frankfurter years earlier as co-founding members of the communist (at that time) American Civil Liberties Union. With so many "civil libertarians" around, one might assume Reich would get an honest break in the legal proceedings, that the judges would actually *read* his documentation on FDA lying, and at minimum throw out the demands for book-burning. And of course, the judges *know* the US Constitution, don't they? But... no. Reich is condemned, and his books are approved for burning.

Reich is imprisoned on 22 March 1957 at Lewisburg Federal Penitentiary. Various psychoanalysts and psychiatrists line up to declare Reich "insane". His time in prison is miserable, he works in the prison library, but is ridiculed as the *"sex-box man"* by other inmates.

Adding insult to injury, the ACLU, whose founding members and fellow travellers played key roles in steering Reich into a prison cell and in the burning of his books, now issues a weak public statement against the FDA book-burning,[161] suggesting feigned alarm for a later claim that they "tried to stop it."

The prisoner occupying the cell adjacent to Reich's, meanwhile, is the schizoid Communist, FBI dis-informant and possible Soviet spy Harvey Job Matusow, who is serving a 5-year sentence for perjury in his role as both claimed witness and assistant to the McCarthy hearings. Matusow obtained a job working for McCarthy, with the goal to introduce false accusations against others, knowing these would later be exposed and hence discredit and destroy the hearings, to make McCarthy look incompetent and ridiculous.[162] The Foreword to Matusow's gloating and untrustworthy book, *False Witness*, was authored by Albert Kahn, another American Communist Party activist and Venona-identified Soviet spy.

Reich dies in prison on November 3 of heart failure, a week before his parole hearing, when he is planning to move to Switzerland with Aurora. His daughter Eva observes Reich's body was abused during autopsy, his innards being kept in a tub and not restored. Reich's last manuscript, entitled *Creation* and written while in prison, and which includes his new *functional orgonometric equations on gravity and negative-g*, vanishes from his prison cell and is lost.[1] Reich's family, friends and close associates gather for his funeral at Orgonon. Multiple malicious obituaries are thereafter published, which repeat the smears and lies by Reich's many detractors. For example: *Time* magazine, 18 November 1957 is merciless, and repeats the original Brady slanders responsible for starting the avalanche of hate and filth against Reich, which led the FDA and US Courts towards his destruction.[63]

TIMELINE POSTSCRIPT: Senator Joe McCarthy, vilified for decades by gulag-denying Marxists and anti-American leftists for his anti-communism, was in later decades proven historically correct about many Soviet spies and untrustworthy persons working within sensitive security positions of the US government.[163] McCarthy was falsely "discredited" by Matusow and other communist apologists in mainstream media (ie, Edward R. Murrow[164]) who inhabited the same left-wing circles that slandered Reich as a routine.

Wilhelm Reich was also eventually proven correct about communists leading the attacks against himself.

Matusow emerged from prison to later take up the freakish role of "Cockyboo the Clown", publicly crowing he was "the guy who

Reich is sent to Lewisburg Federal Penitentiary on 22 March 1957. He dies less than 9 months later.

defeated McCarthy", and also questionably claiming to be a Wilhelm Reich sympathizer.[51,165]

Michael Silvert was released from prison in December 1957, committing suicide six months later.

Mildred Brady was elevated to the post of Editorial Director of *Consumers' Reports* magazine, for the Consumers' Union, which never had a critical word for the FDA or AMA.

The defamation and slander hurled against Reich by his detractors in psychoanalysis and psychiatry, and by smear-writers Brady and Gardner, continue to be repeated, parrot-like, and reverberate within mainstream, academic and pornographic media over the next 60 years.

Gansevoort Incinerator, New York City
Approximately six tons of Reich's books and research journals were seized by FDA agents and burned in New York incinerators on several occasions, starting in 1956. It was *the most outrageous example of Federal government book-burning in American history,* all fully approved by unethical FDA bureaucrats and court judges, including all nine US Supreme Court justices.

3. After Reich's Death: Verification of Reich's Findings by Other Scientists and Physicians, With Renewed Assaults

*"He was a man, take him for all in all.
I shall not look upon his like again."*
Hamlet

Attacks Upon Reich's Research Legacy *Postmortem*

Wilhelm Reich's laboratory and home were closed and shuttered after his death, as issues regarding the trusteeship of his estate were being determined. He left the bulk of his estate to the *Wilhelm Reich Infant Trust Fund*.[1] Some years later, this Trust would open his laboratory to the public as the *Wilhelm Reich Museum*,[1] and begin republishing his books.

Reich's associates were emotionally fractured and scattered after his death. Several of his non-medical supporters soon published scathing accounts of his mistreatment and the illegal, anti-Constitutional abuses he had been subjected to. These included *The Persecution of Wilhelm Reich and the Silence of the Press*[2] and *Wilhelm Reich Memorial Volume*,[3] by Paul and Jean Ritter, *The Murder of Reich, 1957*,[4] by David Boadella, *The Jailing of a Great Scientist in the USA, 1957*,[5] by Lois Wyvell, and *The Ending of Wilhelm Reich's Researches*,[6] by Charles Kelly. Mainstream media and science publications almost systematically refused to publish letters submitted to them decrying the abuses heaped upon Reich over the years – following the same pattern of abusive censorship as the media took during his lifetime, in both Europe and America.

By 1960, Mary Boyd Higgins became Trustee for the Reich estate, and convinced Roger Straus, head of the Farrar, Straus and Giroux publishing house in New York, to republish Reich's major works, starting with a volume of his *Selected Writings*.[7] The FDA

89

learned of this and threatened Straus over it, with the same sort of thuggish, anti-Constitutional push for book-burning as they had done with Reich. Straus stood firm, however, and basically dared them to do so. The FDA backed down, and the first of Reich's many books appeared under their imprint, marking the start of a long series of Wilhelm Reich republications by Farrar, Straus and Giroux. In 1961 they republished Reich's *Function of the Orgasm,* in 1962 *The Sexual Revolution,* in 1963 *Character Analysis,* and so on.[8]

In 1967, the *American College of Orgonomy*[9] was formed by several of Reich's medical associates and trainees, under the leadership of Elsworth Baker, and began publishing the *Journal of Orgonomy.*[9] This new journal carried many original research articles replicating Reich's experiments and showing positive results in favor of Reich. The *Journal of Orgonomy* has been joined over the years by other scientific journals dedicated to a factual recounting of Reich's life and work, such as the *Annals of the Institute for Orgonomic Science,*[10] and this author's own *Pulse of the Planet* journal.[11] These and other journals outside of the English language[12] all shared a common mission to examine Reich's ideas fairly and openly, reporting on clinical case studies of emotion-release therapy sessions and experimental treatments with the orgone accumulator of humans and laboratory mice, including for cancer and other diseases. Other experimental reports were printed in those journals, on Reich's bions, the physics of the orgone accumulator, field work with the cloudbuster device, and Reich's cosmological theories. These new studies, just as those published in Reich's time, have shown moderate to excellent positive results.[12] New books were written as well, by individual scientists and physicians, mostly in the English, German and Italian languages, with translations globally in what can be considered today as a full-scale resurgence of interest in Reich's discoveries.

These new steps were increasingly supplemented by investigations undertaken in American and European universities, also validating Reich's original claims about human sexuality, the bions, bioelectricity and emotions, the orgone accumulator, and even the Reich cloudbuster. Most all of these new experimental investigations, as with the older ones during Reich's lifetime, came from individuals with the MD or PhD or other advanced degree from top mainstream universities. I will discuss them shortly.

Reich nevertheless continued to be mercilessly assaulted after his death, in malicious obituaries and tawdry articles appearing in popular and soft-porn magazines.[13] These articles appeared mostly in the same publications which attacked Reich with slander while he was still alive, and frequently soon after newly-republished editions of his works became available. Like rotting corpses rising from the dead, their headlines seethed with angry poison (frequently masked with "sarcasm"), menacing his memory and legacy, and *always* repeating the 1947 slanders of communists Mildred Brady and Martin Gardner.[13]

- *"Cosmic Life Energy Was Just What the Doctor Ordered: Review of Selected Writings by Wilhelm Reich"* (*New York Times Book Review*, 1960);
- *"Con Men: The Man Who Boxed Sex. Dr. Wilhelm Reich and 'Orgone Energy'"* (*True: The Man's Magazine 1962*);
- *"Dr. Wilhelm Reich and his Magic Box"* (*Nugget 1964*)
- *"Morals: The Second Sexual Revolution"* (*Time 1964*)
- *"The World of Wilhelm Reich"* (*Commentary 1964*).

Only in the case of *Commentary*, was rebuttal allowed in their pages.[13]

Various new books and articles appeared over the latter parts of the 20th and early 21st Century, in "top" mainstream science and leftist news media, such as the *New York Times, Time Magazine* and *London Review of Books*, repeating the same Brady-Gardner slanders and lies about Reich. One pornographic film also caught attention, *WR Mysteries of the Organism*,[14] produced and directed by the Yugoslavian communist Dusan Makavejev, who dragged Reich down into his own mud (see Chapter 6). This film would have passed unnoticed except for how it was promoted as a "documentary", and so people who came to view it and rightfully found its pornography disturbing, wrongly concluded it was an accurate depiction of Reich's ideas. Conservative critics have sometimes reviewed such salacious sexual slanders of Reich, believed they must be true, and then composed their own indignant but misplaced attacks.

This first happened with the finger-wagging moralistic book assaulting Reich in 1955, by the Bible conservative Albert Crombie.[15] The pattern continued into more recent years with the 2008 book by

In Defense of Wilhelm Reich

Daniel Flynn, *A Conservative History of the American Left,*[16] which devoted several pages to slandering Reich with accusations acquired from the same old Red propagandists. It was astonishing to see such hit-pieces written by Communist Party functionaries as Mildred Brady, and published in the KGB-run *New Republic,* being referenced and regurgitated by staunch conservatives such as Flynn, who claims the mantle of Cold War scholar. And he's not alone in this. That example defines part of the modern pattern of attacks against Reich.

In the 1970s, an excellent course on *The Life and Work of Wilhelm Reich* was taught at *New York University* by professors John Bell and Paul Matthews. On one occasion it was attacked and disrupted by radical leftists.[17] A professor was fired in Miami, Florida, for daring to introduce Reich's *Mass Psychology of Fascism* into his course.[17] Other examples could be given, especially after c.1970 when "skeptic groups" such as CSICOP[22] were organized by notable leftist-humanists Martin Gardner and Paul Kurtz, who ramped up the war of eradication of new ideas within medicine and the universities. In such organizations, they played the perfect role as seen in prior years, of Marxist-Soviet operatives partnering with Big Government agencies to obliterate authentic free-market approaches to health care. Kurtz also filled the role of porn-promoter, which may be another reason why the CSICOP "fellows" were so keen to destroy Reich. (See the Appendix.)

Reich's ideas continued to be assaulted in popular magazines and books long after his death, and even in a few "scholarly" publications where one might consider the editors would have some sympathy for the facts. Instead, they all contemptuously repeated the same old sexual slanders from Brady and Gardner, systematically omitting all positive evidence in support of his ideas, and blaming the victim Reich for his own death. These included numerous hit-pieces in the "skeptic club" magazines such as *Skeptical Inquirer* or *Skeptic,* and a few soft-porn publications such as *Penthouse,* but did not end there. They also ranged across the spectrum from popular "news" magazines to supposedly "fact-driven newspapers" and "intellectual-literary" magazines and internet salons, including "peer reviewed" (or "smear reviewed"?) medical-psychiatric and natural scientific journals: *Time* magazine; *New York Times; New York Times Review of Books; Lingua Franca; Salon Magazine; American Thinker; International Social-*

ism; Gestalt Journal; Journal of Psychohistory; Journal of the History of Behavior Sciences; Weatherwise magazine, *Psychology Today; Archives of General Psychiatry; Psychiatry: Interpersonal & Biological Processes; Scientific American;* and most recently *Nature* magazine.

Chapter 5 discusses this last case from *Nature* specifically, where strong action by Reich's supporters finally got them to retract some of their false statements. But this list above only references a small percentage of the published hit-pieces between c.1970 and 2012, beyond those appearing during Reich's lifetime (covered in the Timeline Chapter) or as were published shortly after Reich's death, previously discussed. It does not include an equal number of "academic-literary" books containing sections or chapters repeating the usual slander. *Nor does it include any review of the approximately 25 different post-2011 major newspaper or magazine slander articles which exploded across the mainstream media following the publication of the Turner "Orgasmatron" book.* Those I shall address separately in Chapter 4. The reader must consult the online *Emotional Plague Bibliography,* which lists well over 100 of such smear-slander attack items, for their exact citations.[13]

Unfortunately there does not exist a comparative list of corrective articles, nor even Letters to the Editor where Reich's supporters were given the opportunity for public rebuttal. When rebuttals were written, in most cases they only got private circulation, or were posted to internet in more recent years, as they were refused publication by the original publications printing the slanders. The recurring pattern is: Extensive smears and slanders are written in lengthy articles. The authors and editors then have a good laugh, and copycat articles are stimulated. But when a scientist or physician who knows the facts about Reich's work sends them a serious rebuttal, in 90% of the cases, it is not allowed into print, even in the claimed "scholarly-scientific" journals.

Those interested in Reich's work have been accused of being "obsessed with Reich", but the very long and never-ending stream of attacks and slanders directed *against Reich* and his supporters suggests this is psychological projection. The facts speak more powerfully to *a very real psychological fixation and obsession by his irrational critics, to destroy Reich's research legacy and the facts about his work, with a cold maliciousness lasting now for over 80 years!* This "obsessional stalking" and harassment of the Reichian scientist, physician and educator can be best understood through comparisons to other kinds of deadly stalking of individuals. Consider the example of a deeply armored, sexually impotent male, desperate to love a particular woman who has come to his attention. For emotional structural reasons, however, he cannot make any open and friendly contact – or as in the case where he does, but is rejected by the woman. Due to lack of sufficient emotional or sexual outlets, the obsessive-compulsive man's energy-tension and fixations build to the point where *he begins stalking the woman, but eventually only wants to kill that which now gives him the unbearable excitement*, in a confusion of pleasure and pain, love and hate. Reich identified such obsessive-fixation, aiming to kill and destroy that which an emotionally sick person finds most attractive and exciting, but which for emotional characterological reasons they cannot embrace, as the *Emotional Plague*.[25] It is something most clearly seen among high-energy but emotionally-sexually thwarted individuals, as in the historical attacks by celibate Catholic Inquisitors upon women, who were burned alive, or in modern times by Islamic mullahs and their army of devotees, who attack "sinful" women with acid for not wearing a veil, or stone them to death. Emotional Plague reactions can be seen in individual behavior, as with a murderous stalker where sexual jealously is at work, or in more organized forms as with the Nazi or Comintern obsession to wipe out "evil Jews" or "evil counter-revolutionaries", or the Islamic fanatics who murder "evil nonbelievers" or "sinful women", who are blamed for all their anxieties. Always there is an element of great sexual tension and bottled-up rage and hatred at work.

Especially after the 2011 publication of the Turner book *Adventures in the Orgasmatron* (reviewed in the next Chapter) an entire new raft of smear articles have been printed in major newspapers on both sides of the Atlantic Ocean, slandering Reich mercilessly, and always revealing the sexual anxiety and hatreds of the writers. And following the pattern, these are nearly all coming from self-identified liberals, who are in reality very hard-left people, writing for factually hard-left propaganda publications such as *The Guardian, New York Times, London Review of Books*, and so on.[13] The uniformity of thought in these published attacks were suggestive of totalitarian thinking, a "consensus hostility" against Reich and all his findings, and anyone taking him seriously. It is reminiscent of how Galileo's Inquisitors all "knew for certain" that the Earth was *the Center Of The Universe,* and thereby refused any serious discussion to the contrary, condemning without examination all new scientific findings which might trend in opposing directions.

If you write malicious dirt and lies about Reich, you are sure to find an eager publisher. If you dare to write the facts, as Reich did, or as in this work, only rarely will you find an open-minded editor or publisher. They do exist, thankfully, but are quite rare.[26]

Personal Experiences

I've personally been attacked and slandered both within and outside of the academic Ivory Tower, for daring to undertake research investigating Reich's findings. The worst attacks came when the results of my experimental findings were most significant in support of Reich, and sometimes only because I had mentioned his name in a lecture or publication!

When I was a young university student, my mentors friendly to Reich warned against being too casual in mentioning him, as it might trigger bad reactions from faculty. Indeed on a few occasions during my undergraduate years I was incautious, and professors whom I considered as open-minded turned into hissing snakes upon hearing the name of "Wilhelm Reich". I was very fortunate to later find a graduate research program at the *University of Kansas* (KU) *Geography-Meteorology Department*, where open investigation of Reich's discoveries was welcomed. This was only possible due to the open-minded assistance of Professors Robert Nunley and Robert Haralick. They and a few others believed I should have a fair chance to experimentally investigate Reich's controversial work, to either show there was something to it, or move on to another topic.

My proposal to KU was an experimental test of Reich's *cloudbuster* device, by which rains could be triggered over large areas. That was one of his last discoveries. If proven true, this would have theoretical implications for legitimacy of all of his prior findings, which constituted steps along the path to that final radical discovery on atmospheric processes. I put forth a serious research plan which the Department senior faculty reviewed and critiqued, after which full support was given, equal to that afforded to any other student in the university. While no funds were provided, with help from a friend who owned a metal fabrications shop and by my own resources, I constructed an excellent cloudbuster based upon Reich's

In Defense of Wilhelm Reich

original specifications. The apparatus was set up on the rural farm of Prof. Nunley, who frequently acted as independent observer and even helper in some of the operations. In addition to a full load of graduate-level classes, I acted as teaching or research assistant to various faculty, and at one point was elected by the faculty as the top student in the Department, which provided a number of new opportunities and openings. I completed a two-year field research project investigating the Reich cloudbuster, demonstrating increases in rainfall over the entire State of Kansas following operations. A separate investigation proved it could also dissipate clouds. Pre-operations periods and unaffected clouds were used as controls, with standard National Weather Service data used for evaluation of results. Overall, it was a significant support for Reich, validating one of his most controversial discoveries.[18] I was then awarded the Master's degree, which opened the door for more expanded research and study towards the doctorate.

By that time I was tutoring other students, and was appointed temporary Instructor for large classes in Physical Geography and similar topics. At one point, my classes swelled to over 200 students, held in a large lecture auditorium with three teaching assistants working under my direction. I also proposed and taught several of my own courses, on *Renewable Energy Systems,* and a controversial one, *Rhythms and Cycles in the Natural World.*

I was told no other graduate student in the Department had ever been allowed to take on such responsibilities. Overall, there was an atmosphere of authentic interest in Reich, and at one point I could foresee the development of a natural-scientific *Department of Orgonomic Studies.* Unfortunately, this never came about. But that is how the university system is supposed to function, in support of promising new findings. At Kansas, it did so, swimmingly for a period, in spite of various controversies which invariably developed.

This authentic scientific openness was not found elsewhere in the Academy, however, and a storm was gathering. In a follow-up study, when needing funds to purchase essential research equipment for investigating orgone energy effects upon water, a small grant proposal was drafted and sent to the *National Science Foundation* (NSF) – something that normally is rubber-stamped and automatically approved by NSF, since my home Department also had given its approval. Instead, angry telephone calls condemning my project went from the NSF to the Dean and President of KU, and from there to my Department and graduate mentors. As I was told, *the entire university was at risk of a full cutoff of funds from the NSF, affecting every department in the whole university,* unless they quickly found a way to "stop DeMeo" and get rid of my research program investigating Wilhelm Reich! Against such stiff opposition from "on high", which clearly was some kind of intensive emotional reaction combined with maliciousness, nothing could be done by a vulnerable student, even with support from several tenured professors. Thereafter I did my research on the cloudbuster and orgone energy question more quietly, and in private settings.[19]

At a later point when I was Assistant Professor in the *Geography-Geology Department* at *Illinois State University* (ISU) but finishing up the last bureaucratic details for the final awarding of my Kansas PhD (on the subject of cultural responses to climate change), several KU physics professors wrote angry letters against me, in apparent effort to thwart my doctorate. The physicists were upset because they learned that, about 7 years earlier I had written critically of the Einstein theory of relativity, in an Appendix document to my thesis on the Reich cloudbuster.[20] In that case, my home Department and mentors just ignored the letters of the hypersensitive physicists, and I was awarded the PhD for my extensive research project on the origins of violence in ancient desertification, my *Saharasia* discovery.[21]

More serious, however, were later threats to my professional position at ISU. There, I came under slanderous personal attack by the organized "skeptic clubs", as well as by a few *über-orthodox* professors wanting to "cleanse *their* university" of anyone having unconventional ideas. There was a pattern to the attacks.

A few local academics, well-known as "skeptic" types, made "scholarly" assaults against my work, and my person, both in gossip and letters. These were accompanied by more severe attacks in the "skeptic" press. Then there were dirty-tricks, such as hate-mail, and threats of violence and burglary.[22] These were undertaken by frequently anonymous "footsoldiers" drawn from the "skeptic-club" memberships, or from readers of their publications.[22]

During the 1970s and 1980s, those who were publicly known as authentic Reich scholars or physicians suffered such criminal harassment, which paralleled the publishing of malicious articles against Reich and orgonomy in the "skeptics" press. The homes and laboratories of these workers were burglarized – but only their personal papers and documents were stolen. The *Wilhelm Reich Museum* was burglarized as well, by a pestilent character who was later arrested with stolen items, and identified as a correspondent with the CSICOP "fellow" Martin Gardner.[23]

During my work at ISU, my family, friends and professional associates, the chairmen of universities where I taught, and editors of journals where I published, all received destructive hate mail accusing me of being a "Reichian crackpot" and other slanderous things.[22] At one point, *no less than five chairmen of different physical science departments at ISU wrote letters to the Dean, calling for my head on a platter, and for reasons no better than to merely declare, ex-cathedra, that Reich was a crackpot and nobody engaged in such work should be tolerated.* A new course I proposed on Renewable Energy Systems – covering the principles of energy conservation, solar energy and wind power – was also attacked, notably by departments with heavy funding from the nuclear power vendors. An "ethics committee" was convened at one point, stacked with friends of my attackers, wolves in sheep's cloak who then tried to pull unethical dirty-tricks. (Beware of "ethics committees"!) Thankfully one friendly professor warned me about the situation in advance, thwarting the "attack of the ethicists".

I fought back vigorously with facts and truth. To set the record straight, I wrote rebuttal letters and articles, and provided documentation on the substance of my work and the irrational nature of my critics.[22,23] As I was well-respected in my home Department, where my courses on Earth Science and Climatology attracted large numbers of students, everyone was shocked about the attacks; they had never previously witnessed such social poison being spread around so casually and deceitfully. Over objections by the Dean, my Department sponsored an open lecture to the university faculty and the public, where I detailed my research investigating Reich's findings. This reassured my colleagues, though naturally none of the cowardly attackers bothered to show up. They continued with letter-writing assaults, however, which spilled over into the local newspaper. I then consulted a local attorney. The primary plague characters were warned of a lawsuit if they did not stop with the slander, after which things quieted down. But this unpleasant episode blocked my transfer to a tenure-track appointment, without which I only had an existing limited contract to finish.

I completed three years as Assistant Professor at ISU, after which I had additional similar appointments at the *University of Miami*, and later a brief posting at the *University of Northern Iowa*. Overall I spent a decade teaching and doing my research within the American university system, punctuated by various attacks from skeptic-clubbers and the occasional incensed professor. There were other outrages, such as when a Symposium I had organized to be held at the 1994 annual meeting of the *American Association for the Advancement of Science* (AAAS), entitled *Wilhelm Reich Reconsidered*, was attacked by skeptic-club professors. They did not raise any rational criticism, but instead attacked with jokey but slanderous accusations, such as that "Reich was an antisemite" – which was laughable given his own Jewish heritage, and also given the Jewish backgrounds of several of the proposed speakers. My own presentation was on an historic drought-abatement experiment which had ended years of drought for Israel, with historic rains no less. The AAAS bureaucrats nevertheless unethically dropped the Symposium like a hot potato, in spite of the high level of credentials and serious topics scheduled for presentation.

In another case, a bureaucrat in the Washington DC *Office of Weather Modification* (devoted to cloudseeding) destroyed a file of published papers I had sent to him, documenting and proving the rain-enhancing effects of the cloudbuster. He wrote me a "thank you" letter, but behind my back circulated a lying letter to other government departments claiming there was *no evidence* supporting Reich's atmospheric work, sending that disinformation also to the Agricultural Ministries of several dryland nations with whom I had been organizing new research projects – resulting in cancellation of those projects. In another such example I approached the *Drought Emergency Task Force* in California during very severe drought conditions which had caused billions in crop damages, offering to use the cloudbuster in a demonstration project that would not cost them a penny. I got no response from them, but a few days later a ridicule slander article appeared in the newspaper of the hardest-hit drought region, attacking me and Reich, thereby muddying the waters so badly it became impossible to do anything.

And on and on it went.

After 1988, I tired of bouncing around the country for limited university appointments, a situation created by chronic attacks by the "skeptics", and focused my efforts on building up a private research institution. With considerable work, I now have established the *Orgone Biophysical Research Lab*[24] in rural Oregon. Since "going private", I have received more invitations to speak at universities, and requests at the "official level" to undertake field work with the cloudbuster, to end serious droughts, than previously when I held a professor's position. But the sad fact is, in those nations or regions where the academics, politicians and general populations have embraced such slander-writers and media liars, they have lost all the benefits from Reich's discoveries, as well as benefits from other new discoveries that would have been treated with similar abuse, for similar irrational reasons.

Overall, these experiences were a stark lesson in emotional plague politics, and also on something Reich once complained about, *the high-minded academic or "civil liberties" journalist who speaks so glowingly about "academic freedom", "freedom of inquiry" and "freedom of speech", which they claim for themselves and their friends, but have no hesitation to deny to disliked others when it suits them.* And on this score, I should note that, contrary to my early expectations, it was primarily *ordinary conservatives* who rallied to my defense, to my rights to undertake controversial work, to support my work, and to defend my rights to freedom of inquiry. *Virtually ALL of my attackers were either Marxist or left-liberals of one type or another.* This was initially quite confusing given my own "old fashioned" liberal tendencies. I expected the liberals to rally to my side, to defend basic academic freedom, and to appreciate Reich's work on sexual freedom and new energy ideas. After all, a primary liberal sentiment I had been brought up with was the importance of freedom of speech and of the press, so one imagined liberals would become quite incensed about book-burning. As Voltaire, Burke, Paine, and many others stated: *"I may disagree with what you say, but will fight to the death to defend your right to say it".* Occasionally my liberal contacts did express such sentiments, but mostly *not,* and all too frequently went silent or *led the attacks* against my work, with sharpened knives for back-stabbing.

Since this is a book about Reich, I won't say more about my personal difficulties. Under the best of circumstances, when it comes to the heretic Wilhelm Reich, there are *no guarantees* people will respond with fairness or rationality. Let us now discuss some additional supporting evidence favoring Reich's findings and theories, developed in the years after his death.

Recent Independent Confirmations of Reich's Discoveries

Reich's work crossed many scientific disciplines and encompassed so many subjects – moving across medicine, biology, microbiology, physics, cosmology and atmospheric science – that if it was all correct and true, it would have been suitable to earn him three or four separate Nobel Prizes. But is it true? Was Reich correct on this amazing sweep of research and discovery? Based upon my own personal investigations, I will firstly say the answer is *Yes,* and have no hesitation to publicly *stand up for Reich* on this matter.[26] And I do so not by some secondhand "appeal to authority" of others, but in my own role as a field and laboratory scientist, a former university professor and classically-trained natural scientist who has made extensive and firsthand experimental investigations into Reich's central findings, with confirming experimental results. Secondly, I point to the similar authority of many other research scientists and physicians going back more than 70 years, who have likewise tested and verified Reich's findings, and who also have stood up for Reich.[12,26]

Most all of these confirming studies were undertaken by individuals with MD or PhD degrees from mainstream universities, and using the best-available methods and laboratory equipment for their time. Working in private or university laboratories, they employed tightly-controlled procedures, applying systematic and often blinded methodology. Let's review some of them.

Confirmations of Reich's Sex-Economic Findings

One of my earliest and most central research projects was a systematic test of Reich's sex-economic theory on the origins of human irrationalism and social violence. My work on this subject constituted what was (and probably still is) *the largest global cross-cultural review of human behavior and social institutions ever undertaken.* My investigations, now published in the book *Saharasia,*[20] involved a *quadruple-blind* systematic global review[27] of human family and sexual life, and tendencies towards social violence, in over 1100 separate human cultures. Each of these cultures were additionally pinpointed on world maps to identify their regional similarities or differences. I produced the first-ever *world maps of human behavior and social institutions.*

In Defense of Wilhelm Reich

To wit: Human violence originates and *exists most strongly* within cultures having the most intensive expressions of the following two major identifiable characteristics:

1) They abuse their infants and children with heavy-handed socially-demanded rituals that consciously or unconsciously inflict severe painful trauma (ie, genital mutilations), or otherwise to destroy the maternal-infant bond (ie, prolonged swaddling), establishing male dominance over female reproductive functions (ie, anti-contraception, male-dominated obstetrics versus midwifery).

2) They also crush down heterosexual love by punitively obliterating the natural heterosexual interests of the child and unmarried adolescent (ie, Romeo and Juliet), with, in the most extreme cases, severe punishments or the death penalty – to include sexual taboos, demands for virginal sexual ignorance, compulsive lifelong and frequently arranged loveless marriages, vaginal blood taboos, prohibiting women the rights to control their own fertility and life-destiny – and also by simultaneously permitting adult molestation and rape of emotionally-crushed children and adolescents, notably by pedophile attacks on boys, and arranged marriages of girl children to old men.

The more a culture engages in such practices, as my work demonstrated, the greater was its quanta of bottled-up emotional misery and rage. Out of this comes their higher levels of social violence, and the tendency to engage in aggressive wars and atrocities against other cultures or minorities and social dissenters within their own society.

These findings confirmed Reich exactly!

I can also mention the prior work of neurophysiologist James Prescott who also undertook a cross-cultural investigation entitled *Body Pleasure and the Origins of Violence*.[28] Prescott used a smaller 400-culture sample of the tabular cross-cultural data as later used in my 1170-culture Saharasia study, verifying a similar social-emotional pattern in the sex-repressive and child-abusive origins of human violence. These cross-cultural studies stand as proof-positive, beyond Reich's original clinical findings and social observations, that *human social violence and war are socially and environmentally determined, and not something innate or genetic.*

In addition to these studies, significant clinical work has been undertaken by therapists applying Reich's psychotherapeutic methods, to help people overcome neurotic compulsions or inhibitions, and more severe problems of emotional/mental illness.[12] Reich's clinical research findings, and his larger concepts of genitality and emotional energy as foundations for mental and physical health, are more specifically detailed in works by the physicians Elsworth Baker in *Man in the Trap*,[29] Morton Herskowitz in *Emotional Armoring*,[30] and Stefan Müschenich in *Der Gesundheitsbegriff im Werk des Arztes Wilhelm Reich (The Concept of Health in the Works of Dr. Wilhelm Reich).*[31]

At the time when Reich did his original work on the subject of human sexuality and orgastic functioning, he attracted malicious opposition for merely having the words "sex" or "orgasm" in the title of his books. Today, by contrast, there are too many such books to even briefly list, which indicates a greater openness towards sexual topics. However, the general theme of most of such books reaffirms Reich's SexPol observations made decades ago, of very high levels of orgastic impotence, sexual dissatisfaction and unhappiness. The high levels of divorce are also a testament to this fact. Very few of the modern "sex researchers" acknowledge Reich's priority on the scientific or clinical aspects of sexuality; unfortunately, the Kinsey agenda is central to most of them. (See the Kinsey section in the next Chapter.) So what was identified as neurotic or psycho-pathological in Reich's work is today increasingly "normalized". The reasons for this are too complex for any discussion here, but suffice it to say *this is due to a retreat by scientific professionals in the face of frequently violent political activist sex-identity groups*, not because scientific investigation has affirmed Kinsey or refuted Reich.

Confirmations of Reich's Bioelectric Experiments

Reich's early work at the University of Oslo demonstrated the existence of specific and objective bioelectric phenomena which appear during subjective human sexual and emotional excitation.[32] His work on this subject clarified the *basic antithesis between the sympathetic and parasympathetic nervous systems.* With subjective feelings of pleasure and expansive relaxation, which are parasympathetic nerve functions, bioelectric charge at the skin surface increases. Sympathetic nervous reactions, associated with anxiety and unpleasure, show a loss of bioelectrical charge at the skin surface. Reich's experiments constituted the first time subjective perceptions had obtained objective biological measures in a laboratory. This clarified many open questions in neuroscience about how traumatic experiences could elicit nervous and muscular reactions tending towards chronic anxiety.

Reich's work on bioelectricity gained indirect confirmation via the later works of experimentalists such as Harold Burr (writing on *electrodynamic fields*),[33] Robert O. Becker (writing on *The Body Electric*),[34] and Bjorn Nordenstrom (writing on *Biologically Closed Electric Circuits*).[35] All of these research scientists identified bioelectric potentials which are determinants of human growth and functioning, and of health or disease, in keeping with Reich's view that bioelectric charge is an expression of a fundamental life-energetic phenomenon.

Reich's more exact experimental approach on bioelectric investigation of sexuality versus anxiety, underlain by a pulsatory flow of bioenergy for which bioelectric potentials were a secondary expression, was only systematically verified in more recent years, by physicians Byron Braid and Robert Dew (1988),[36] and by Günter Hebenstreit at the University of Vienna (1999).[37]

Confirmations of Reich's Bion Experiments

The *bion* is a transitional microbial form of around 1 micron diameter and visible in the light microscope, discovered by Reich during his Norwegian period of research.[38] They are created from both living or nonliving matter through processes of swelling and disintegration in water, and speeded in this process by heating material to red-hot incandescence before immersion in sterile nutrient solutions, as well as by the freeze-thaw process. Bions exist between the worlds of living and nonliving matter, with relations to what are today called *Archaea* or *extremophiles*. They provide an answer to many mysteries in biology, as well as a solution to the *Origin of Life* question.

Reich's bion experiments are explicitly described in his research journals,[12] and in *The Bion Experiments: On the Origin of Life*.[38] They are nevertheless hotly controversial, and were maliciously attacked during his European period of research, as already described. But the bion experiments have been repeatedly replicated by biologists such as Bernard Grad of McGill University, and many others. The verification studies are in fact too numerous to list here, so I must once more refer the reader to the aforementioned online

Bibliography on Orgonomy,[12,§] where a browser term-search for "bions" will turn up the citations to many published accounts, several of which are also available online and provide color images. Reich's work on this question predates some of the better-known investigations on the lifelike particles and life-forms as found much later in *Martian meteorites* and *deep sea hydrothermal vents*, but he is rarely if ever cited for his priority.

I've also undertaken replications of Reich's basic bion experiments. In October 2000, I presented a confirmation of Reich on this subject to the *Second International Symposia on Pleomorphic Microbes in Health and Disease,* in a presentation now available as both a published paper and as a YouTube video.[39] My Institute hosted laboratory seminars on this subject for about 10 years starting in 1996, variously led by physicians Richard Blasband and Stephen Nagy, or biologist Bernard Grad.

These findings on the bions also underscore Reich's important discoveries on the *origins of cancer and the cancer cell,*[40] for which science, medicine and humanity owe him a great debt. I'm not a physician and so do not treat people, but others have applied Reich's methods, and have gone the additional step to verify his findings on the *orgone energy accumulator* as a powerful therapeutic device.

Confirmations of Reich's Orgone Energy Accumulator
A. The Biological Effects

Reich's orgone accumulator was subjected to a double-blind and controlled study of human physiological response in a 1986 study undertaken by Stefan Müschenich and Rainer Gebauer at the *University of Marburg* in Germany.[41] This study demonstrated statistically-significant biological effects upon heart-rate, blood pressure, core body temperature and other parameters within a group of volunteer subjects who knew nothing about the accumulator. The effects vanished and were not observed when volunteers sat in a nearly identical but non-orgone-accumulating dummy box. A similar controlled-blinded verification study was undertaken in 1995 by Günter Hebenstreit at the *University of Vienna*.[42] All of those studies as well as many clinical case-studies have been

§ A new book discussing the bions has recently been published: Peter Jones: *Artificers of Fraud: The Origins of Life and Scientific Deception*, Orgonomy UK 2013.

In Defense of Wilhelm Reich

published over the years,[12] exactly verifying Reich's documented accounts on the biological effects of the orgone accumulator.

In my own studies of Reich's work, I had the occasion to build many different orgone accumulators, and so could personally verify the subjective impressions and general energy-boosting capacities of the device. But it was frustrating that there was no way for most people to easily learn about Reich's discovery and apply the accumulator for their health problems. None of the physicians interested in Reich whom I knew were *openly* treating people with the accumulator, out of fear of the FDA or loss-of-license from State medical boards and such. By the late 1980s, however, having seen several of my loved relatives pass away from cancer (or due to the heroic/horrific cancer treatments) I wrote a book on this subject: *The Orgone Accumulator Handbook.*[43]

My *Handbook* gives basic plans for proper construction of orgone accumulators and blankets, and is a self-help for people wanting to supplement or break free of the allopathic MD-hospital system. It was also written at the time of repeated FDA raids upon natural healing clinics, so as to insure Reich's discoveries on the orgone accumulator could never again be wiped out by power-drunk government officials or anti-health medical societies. It gives a listing of various research papers on the clinical use of the orgone accumulator against various diseases and conditions. I will quote from an essential part of the 2010 edition of my *Handbook,* and again refer the curious reader to the online Bibliography[12] for the exact citations to what is identified herein:

[Quote][43: p.134-141]
Clinical Case Studies of Disease Treatment

Here is a listing of published papers giving the diseases treated along with the name of the physician who authored the paper, and the year of the publication.[12]

Disease/Condition	Physician/Author	Year
Cancer Biopathy	Wilhelm Reich, MD	1943-48
Cancer, Burns	Walter Hoppe, MD	1945
Mediastinal Malignancy	Simeon Tropp, MD	1949
Multiple conditions	Walter Hoppe, MD	1950
Multiple conditions	Victor Sobey, MD	1950

Rheumatic Fever	William Anderson, MD	1950
Breast Cancer	Simeon Tropp, MD	1950
Ichthyosis	Alan Cott, MD	1951
Manic Depression	Philip Gold, MD	1951
Hypertensive Biopathy	Emanuel Levine, MD	1951
Leukemia	Wilhelm Reich, MD	1951
Cancer	Simeon Tropp, MD	1951
Diabetes	N. Weverick, MD	1951
Coronary Occlusion	Emanuel Levine, MD	1952
Multiple conditions	Kenneth Bremer, MD	1953
Skin Cancer	Walter Hoppe, MD	1955
Pulmonary Tuberculosis	Victor Sobey, MD	1955
Uterine Cancer	Eva Reich, MD, W. Reich, MD	1955
Uterine Cancer	Chester Raphael, MD	1956
Rheumatoid Arthritis	Victor Sobey, MD	1956
Malignant Melanoma	Walter Hoppe, MD	1968
Cancer Biopathy	Richard Blasband, MD	1975
Cancer Biopathy	Robert Dew, MD	1981
Multiple conditions	Dorothea Fuckert, MD	1989
Skin Infections	Myron Brenner, MD	1991
Cancer Biopathy	Heiko Lassek, Artz.	1991
Multiple conditions	Jorgos Kavouras, MD	2005

<u>Controlled Studies with Laboratory Mice</u>

There also are many controlled experimental studies undertaken with laboratory mice, evaluating the effects of the orgone accumulator or orgone medical dor-buster (a related apparatus) upon their health and longevity. This includes mice, either genetically predisposed to develop spontaneous tumors or leukemia, or who were given tumor transplants. As noted, these studies showed considerable improvement in the health of these immunologically-stressed or weakened mice when given a daily treatment with the orgone accumulator, as compared to otherwise identically-treated control groups. This was reflected in their overall descriptions and vitality factors as detailed in the various papers, but mostly is objectified in their dramatically increased life-span. Orgone accumulator treatment increased the mouse life-span from 1.6 to 3 times as long as the controls!

In Defense of Wilhelm Reich

1. Wilhelm Reich: "Orgone Therapy Experiments", in *The Cancer Biopathy*, Orgone Institute Press, Rangeley, ME 1948 (Farrar, Straus & Giroux, 1973, p.290-309). ...**The orgone accumulator approximately tripled the life-spans of the treated mice.**

2. Blasband, Richard A.: "The Orgone Energy Accumulator in the Treatment of Cancer Mice", *Journal of Orgonomy,* 7(1):81-85, 1973. ...**The orgone accumulator treated group lived 1.6 times as long.**

3. Blasband, Richard A.: "Effects of the Orac on Cancer in Mice: Three Experiments", *Journal of Orgonomy,* 18(2):202-211, 1984. ...**The orgone accumulator treated mice, developing spontaneous tumors and treated early, lived nearly twice as long.**

4. Trotta, E.E. & Marer, E.: "The Orgonotic Treatment of Transplanted Tumors and Associated Immune Functions", *Journal of Orgonomy,* 24(1):39-44, 1990. ...**The orgone accumulator more than doubled the life-span of the treated mice.**

5. Blasband, Richard A.: "The Medical DOR-Buster in the Treatment of Cancer Mice", *Journal of Orgonomy,* 8(2):173-180, 1974. ...**Medical dor-buster treatment alone led to a significant 50%+ increase in longevity.**

6. Grad, Bernard: "The Accumulator Effect on Leukemia Mice", *Journal of Orgonomy,* 26(2):199-218, 1992. ...**The orgone accumulator treatment reduced their leukemia incidence by around 20%**

7. Baker, Courtney F., et al: "Wound Healing in Mice, Part I", *Annals, Inst. Orgonomic Science,* 1(1):12-23, 1984. "... Part II", *Annals, Inst. Orgonomic Science,* 2(1):7-24, 1985. ...**Observed increased healing by orgone accumulator treatment from a nominal 1% to 12% increase in Therapeutic Index, and were statistically significant.**

Conclusions: Overall, these studies indicate *the orgone accumulator is most effectively beneficial when applied quickly after identification of the illness or injury. The most reproducible anticancer effects in laboratory mice were observed primarily where spontaneous tumor development occurred. A lesser, but noticeable and important anticancer effect was observed in the case of transplanted tumors.* This is in keeping with observations from published clinical case-studies of orgone accumulator therapy on human cancer patients.

The reader might correctly complain there are only a few biological studies to show after so many years following Reich's death. However, one must appreciate how all of these physicians and scientists took great personal and professional risks in doing this kind of research. The chronic open warfare against orgonomy by the FDA and medical groups, which has existed since the 1940s, has taken its toll. Nevertheless, ***everything here confirms Wilhelm Reich's original positions, and powerfully suggests the orgone accumulator ought to be available for use in every home, clinic and hospital, worldwide.***
[End Quote] [43: p.134-141]

One of the physicians mentioned in the above section has written a book on his experiences in successful applications of the orgone energy accumulator to patients suffering from a variety of disease symptoms, over years of work in a rural German clinic: *Heilen mit Orgonenergie: Die medizinische Orgonomie,* by Jorgos Kavouras. [44] And he is not alone in this. [12] This work was only possible in Germany because *German Law today acknowledges the central role of the medical profession in creating and maintaining the Holocaust death-camps.* [45] We in America ignore this lesson from history, and empower our hospital-MDs and FDA, which burns books and imprisons pioneering physicians and health scientists. [46]

The orgone accumulator has also been experimentally investigated from the viewpoint of Chinese medicine and acupuncture. Studies have directly proven that *orgone energy is the objective phenomenon which Chinese and Hindu culture had previously identified (and frequently mystified) as Chi or Prana.* Specialized small hollow orgone accumulators made in narrow tube-form, can elicit reactions in the body identical to acupuncture needles,[47] but without any direct touching or contact to the skin surface. Other studies have shown positive physiological reactions of patients to the orgone accumulator using bioelectrical *Ryodoraku* instrumentation, and other measures.[48]

Numerous studies have also been published on the effects of the orgone accumulator on plant growth. I undertook a multi-year study on the effects of the orgone accumulator upon sprouting mung beans.[49] A systematic 34% boost to growth of the beans was observed when sprouted inside the orgone accumulator during *optimal orgonotic conditions*,[50] as compared to a control enclosure kept under nearly identical temperature, humidity and light conditions. In that paper I surveyed other studies which, while not so systematically undertaken, showed similar results. This particular experiment was performed annually by the students in my seminars, to convince themselves of the reality of orgone accumulator influences.

Confirmations of Reich's Orgone Energy Accumulator
B. The Physical Effects

Replication studies have also been made, showing excellent results, for Reich's findings on anomalous physical phenomena within the orgone accumulator. Nearly all the major anomalies identified by Reich have since been confirmed by others, with new ones uncovered, using even more precise measuring and recording equipment than was available in Reich's day. As before, the number of papers on this subject are simply too numerous to list, so again the reader is referred to the online *Bibliography*[12] for a complete listing.

The physical anomalies inside the orgone accumulator include such things as a spontaneous slight thermal warming, the slowing of the natural electroscopical discharge rate (*natural leak*), and anomalous behavior of orgone-charged Geiger-Müller ionization-type radiation detection tubes.[12,51,52]

I've personally undertaken most of these experiments, including a tightly-controlled multi-year study verifying the *thermal anomaly within the accumulator*.[53] A pulsatory slight thermal heating spontaneously builds up inside the orgone accumulator of around 0.1°C. average, pulsating with peaks of up to 0.7°C on days of lowest humidity and clear weather, when the accumulator is known to produce the strongest charge. Atmospheric water binds up the local orgone energy charge, reducing all the anomalous physical and biological effects, especially as humidity climbs above 50%.

Water itself can also be more directly influenced by charging inside the orgone accumulator, in a manner tied to this same humidity factor. Specifically, water evaporation rates from open petri dishes is suppressed inside the orgone accumulator as compared to a control enclosure, but only on low-humidity days when orgone charge in the local atmosphere is strong.[54] Water also shows an increased absorption in the UV frequencies between 240-280nm when charged up inside an orgone accumulator, as compared to control samples.[51,54]

I've also verified Reich on the existence of anomalous high counts-per-minute (cpm) recorded within orgone-charged Geiger-Müller radiation detector tubes. One of my specialized orgone-charged radiation detectors has yielded *~50 to 4000 counts per minute*, while an identical control instrument, not orgone-charged, yields ~10 counts per minute at most.[51,54]

Roberto Maglione also recently compiled his own experimental investigations into Reich's orgone energy and the accumulator, as well as the findings of others, in the book *Methods and Procedures in Biophysical Orgonometry*.[52]

Confirmations of Reich's Atmospheric Discoveries

The physical anomalies discovered in Reich's laboratory work brought his science out into the open field of weather research. One of his most "unbelievable" devices came as a consequence of this line of investigation: the *Reich Cloudbuster*.§

In the 1970s, I studied with Robert Morris, Richard Blasband, and some of the other physicians and scientists who had personally investigated Reich's major discoveries, including the cloudbuster apparatus. I built several orgone accumulators following Reich's designs, and investigated its many biological and physical-atmospheric parameters. This work proceeded parallel to my university studies, as previously described, where I made open investigation of Reich's cloudbuster device with oversight of several senior professors and Reich scholars.

---

§ Not to be confused with the many gadgets misidentified by lay enthusiasts as "cloudbusters" or "chembusters", as on ebay and internet. See the Chapter on *Orgone Flapdoodle* for details.

In Defense of Wilhelm Reich

The University of Kansas investigations yielded positive results in keeping with Reich's original claims, and was something of a small breakthrough – though it also created a great controversy. That study is now published in a book bearing the same title as the original thesis: *Preliminary Analysis of Changes in Kansas Weather Coincidental to Experimental Operations with a Reich Cloudbuster.*[18] My later field work (1980-1995) to end droughts using Reich's methods were undertaken at many different places in the USA.[55] Later still I undertook drought-abatement and desert-greening work in the American Southwest (1989),[56] Israel (1992),[57] Namibia (1994),[58] and a 5-year project in Eritrea, Africa (2002),[59] with official-level support from the respective governments. Entire teams of volunteer associates helped with the actual field work, in cooperative association. All of those projects were evaluated using official weather records showing positive rainfall increases, in keeping with Reich's original findings. Detailed reports were written up, published, and circulated to various government offices where they were typically promptly ignored. My results thereby mirrored and supported not just Reich, but the many prior cloudbusting studies undertaken since Reich's death, as by Blasband, Eden, Baker, and others.[12] A publication by Maglione, *Wilhelm Reich and the Healing of Atmospheres: Modern Techniques for the Abatement of Desertification*[60] provides a complete survey of work done in this field during and since Reich's time.

The above studies have been published in scientific books and research journals, most of them meeting the standards for peer-review of their period (being critically reviewed and overseen by other knowledgable and credentialed scientific people in similar fields of study). Some of them are found within the academic mainstream as well, including as university theses and dissertations which typically receive a far greater critical peer review than the average article in a scientific research journal.

Other Recent Publications on Reich and His Orgonomic Science

To the above, we may add the following short list of additional books on the subject, restricting the list to only those works in English and German, the two languages Reich wrote in.

There are at least five major English-language biographies of Reich since around 1970, written by American, British or

Scandinavian scholars, to wit:

- *Fury on Earth: A Biography of Wilhelm Reich,*[61]
    by Myron Sharaf, PhD
- *Wilhelm Reich: The Evolution of His Work,*[62]
    by David Boadella
- *Wilhelm Reich and Orgonomy,*[63] by Ola Raknes, PhD
- *Wilhelm Reich Versus the USA,*[64]
    by Jerome Greenfield, PhD
- *Wilhelm Reich and the Cold War,*[65] by Jim Martin
- *Wilhelm Reich: A Personal Biography,*[66]
    by Ilse Ollendorff Reich

Additionally several compendiums have been published, of multiple articles by many different professionals, which cover Reich's sex-economic, therapeutic and medical work, as well as his orgone biophysics:

- *Nach Reich: Neue Forschungen zur Orgonomie.*[67]
    edited by James DeMeo, PhD & Prof.Dr. Bernd Senf
- *Heretic's Notebook: Emotions, Protocells, Ether-Drift and Cosmic Life Energy,*[68] edited by James DeMeo, PhD
- *On Wilhelm Reich and Orgonomy,*[69]
    edited by James DeMeo, PhD
- *In the Wake of Reich,*[70] edited by David Boadella

These various works are all in addition to the very important and substantial materials as originally published by Reich himself, as well as all the other journal articles and reports published since Reich's death. They are all separately indexed in the online *Bibliography on Orgonomy.*[12]

Finally I should also mention there have been a half-dozen major studies where the internal documents of the FDA and FBI on Wilhelm Reich (and sometimes, of his detractors) were obtained, via *Freedom of Information Act* searches of government files. These studies are listed at the end of the Reference section, under *Additional Materials for Consultation.* Notable are the studies by Blasband and Baker from 1972-1973, Greenfield from 1983, my own "Postscript" study from 1989, and the Martin study from 2000. They all confirm the scheming and incompetence displayed by the FDA during their "investigation" of Reich. Particularly the study undertaken by Martin reveals a great deal about Reich's detractors and their motivations, and overall supports Reich's own analysis as published in the last years of his life.

These findings by multiple scientific investigators on several continents, and published in research journals and books over many years, have confirmed Reich exactly on all the critical points, notably regarding the existence of a powerful unusual energetic force within nature, which can be amplified within the interior structure of his orgone energy accumulators. But do not try to find any mention of these studies, even one single reference or citation, in any of the ugly screeds written against Reich by his detractors. They don't exist.

There also were several film documentaries produced after Reich's death, with variable degrees of accuracy. The most accurate and educational so far is *Man's Right to Know*,[71] written and directed by Kevin Hinchey, and which is available for purchase from the *Wilhelm Reich Museum*,[1] but also available on YouTube. A moving short dramatic production was developed by filmmaker Jon East, *It Can Be Done.*[72] Other interview-type documentaries exist, such as *Viva Little Man*[73] by Digne Marcovicz, and *Who Is Afraid of Wilhelm Reich*,[74] by Antonin Svoboda. While I can quibble about this or that in some of them, they all present Reich in a sympathetic light,[75] and collectively stand against the worst of the lies that are still carried forward by his detractors in the mainstream media, the FDA and medical community. Reich's critics have not yet given up the effort to slander and defame him into extinction.

Now, let's proceed to review specifically what a few of Reich's worst detractors have produced, in comparison to Reich's authentic biography and research history.

What have Reich's critics written, what have they done with all the information available to them on these large and important subjects?

The answer, unfortunately, is *nothing*. Nothing except to spread *poisonous fabrications, half-truths and lies of omission* about Reich's life and works, and about the later studies. They pretend there is no scientific validity to any of it, no professional support or confirmation, in hopes that nobody will learn the facts.

4. Critical Review *
of Christopher Turner's
Adventures in the Orgasmatron **

"A lie gets halfway around the world before the truth
has a chance to get its pants on."
Winston Churchill

Introduction

British author Christopher Turner, whose book *Adventures in the Orgasmatron* will be the subject of this chapter, adds considerably to the *80-Years' War* of anti-Reich slander and defamation. His various distorted and smutty writings against Reich constitute a new offensive battle, repeating old lies from Reich's European and American detractors of c.1930-1960, along with new ones. For those who have seriously studied Reich from original sources, Turner's errors of fact are fairly obvious and easy enough to refute. The problem with it lays in how uninformed people will stumble over his highly promoted book and be fooled into believing it is authentic biography. In fact, the "Orgasmatron" book assaults, slanders and dismisses Reich as an irrelevant, insane, and even pornographic crackpot. It is also filled with gaping holes from the numerous *lies of omission* (the worst kind of lie, as George Orwell noted). Few words of support are provided for the poor Reich, locked in the stocks, with no serious presentation of rescuing facts from other scientists who have studied his work in depth, nor any other opinions from his defenders, dead or alive. This is a most critical point, as verification studies are a central component of the scientific reception process, whereby new claims are evaluated empirically

---

* Originally web-posted in mid-2011, here:
    http://www.orgonelab.org/DeMeoReviewOfTurner.htm
** Turner, C.: *Adventures in the Orgasmatron: How the Sexual Revolution Came to America*, Farrar, Straus & Giroux, NY, 2011. UK edition subtitled: *Wilhelm Reich and the Invention of Sex: How Renegade Europeans Conceived the American Sexual Revolution and Gave Birth to the Permissive Society*, Harper Collins UK (Fourth Estate).

In Defense of Wilhelm Reich

and then either validated as correct and accurate, or refuted with a pointing to specific errors or flaws in an investigator's procedures or conclusions. So naturally, Turner's book carries *not one authentic reference to any of the many published experimental studies and clinical reports by the scientists and physicians (as in the preceding Chapters), who have independently verified Reich's original findings.*

Turner knows about the large body of confirming work supporting Reich, because he repeatedly met with and interviewed people intimate with his research findings. He cited various authors who are friendly to Reich, even mentioning the *Journal of Orgonomy* occasionally where it could suit his agendas.

Most tellingly, however, Turner was personally present at two conferences devoted to scientific confirmations of Reich's work. He was observed to attend the 2004 Summer Conference at the Wilhelm Reich Museum,[1] on *The Orgone Energy Accumulator: Its Medical and Scientific Use* wherein physicians from Germany and America reported their successful treatments of various disorders and diseases using the orgone accumulator. He also attended the *2007 International Conference on Orgonomy: Wilhelm Reich in the 21st Century,* also organized by the Wilhelm Reich Museum[1] in honor of the 50th anniversary of Reich's death.

This latter Conference hosted a group of around a dozen natural scientists, historians and physicians who reported on their various research efforts, including systematic, controlled and blinded experiments undertaken at major universities which confirmed Reich's science on major points. The presenters discussed subjects such as orgone therapy, replication of the bion experiments, Reich's work at the *University of Oslo,* human physiological reactions to the orgone accumulator, field experiments with the Reich cloudbuster, and so forth. Therefore it is well known that Mr. Turner was exposed to some very good experimental findings in support of Reich's original discoveries. But *not one mention of these supportive findings were included in his book, nor in any of his related articles.* These omissions were a primary breach of professional ethics, compounded by the inclusion of many false claims and half-truths in the book.

In this *Review*, I'll provide short quotes from Turner which will be compared to what Reich and his associates have actually accomplished and written, with added commentary. Detail will also be provided on the large differences between the sexual theories of Alfred Kinsey and Wilhelm Reich, the former of whom appears to figure prominently in Turner's *weltanschauung*. Turner's own writings suggest a celebratory defense of Kinsey's sexually indiscriminate, poly-sex and pedophilic "anything goes" promiscuity. And as I will show, this helps to understand the smothered anger and antipathy (frequently expressed sideways as snide sarcasm) expressed by Reich's detractors towards his findings on *heterosexual genitality*, which are a decidedly moderate and healthy understanding of human sexuality.

The Turner book was promoted as an "authoritative biography" on Reich's life and work. However, it doesn't provide even a superficial accounting of the man or his science. Turner's distortions are extreme with regard to Reich's sexual and biophysical research, the latter of which is hardly mentioned except in short sentences mixed with contemptuous ridicule. This is an old trick, to obscure or omit accurate details while simultaneously putting up straw-men who can easily be knocked around, making them sound fully "unbelievable" to the average person who has no accurate knowledge on the subject. Similar ridicule was hurled against Goddard's rocket and the Wright brothers' airplane, we may note, both of which suffered a deadly retardation in development within the American homeland where they were first invented and developed. For the American public which has witnessed inordinate amounts of deliberate lying on the part of the mainstream media on political matters, they should consider how this trend of open mainstream deceit has for a long time also worked to destroy maverick scientists and biomedical pioneers. This trend began in the early 1900s, when the *American Medical Association* (AMA) made a concerted political push to gain monopoly powers for its MD membership. After c.1938, the FDA was given unprecedented police powers to assist that same agenda. Reich became a victim of that push, though the great anxiety stirred up by his earlier writings on human sexuality and his criticisms of Freudian dogma and Soviet Communism were the more fundamental reasons, as discussed in prior chapters.

Author Turner represents a modern expression of that coalescence of trends, of Big Medicine, Big Media and Big Government coming together to jazz up "consensus outrage", through clever disinformation, against the unorthodox scientist and physician. The "skeptic clubs" are also leaders of that trend, though all of these institutions accomplish their goal of crushing the heretic by inflaming the bottled-up emotional tensions within the average person, in the

In Defense of Wilhelm Reich

manner of a social arsonist.

Turner has played a similar role towards Reich's work. He spent considerable time in both England and the United States, surveying Reich's biography and accomplishments, reviewing materials in various government and university archives, as well as reading books and conducting interviews. *But what has he produced with all the information available to him on this large and important subject?*

Turner's 2004 Swipe at Reich, Orgonomy and Summerhill School

In 2004, Christopher Turner announced his presence with an ugly little article, *"Summerhill School and the Real Orgasmatron"*, appearing in the left-wing *London Review of Books (LRB)*.[2] In it, he denounced the British *Summerhill School,* founded by the late A.S. Neill who was a personal friend of Wilhelm Reich. Reich was the main target of the article, however, being subjected to sexual slander in relation to an old orgone energy accumulator which Turner found on the Summerhill campus. Turner wrote:

> *"While I was there, I discovered that an Orgone Energy Accumulator had once been used at the school... for the purpose of improving its users' 'orgastic potency' and by extension their general, and above all mental, health. It became fashionable in the 1950s and 1960s, and was used by such countercultural figureheads as Ginsberg, Kerouac and Burroughs – who claimed to have had a spontaneous orgasm in his. Woody Allen parodied it as the 'Orgasmatron' in Sleeper. I wondered what place such a machine would have in a school... rearing a child in a cupboard, away from civilisation?"*[2]

In these few short sentences, Mr. Turner gave the airs of a "pop-culture sexual expert", name-dropping and "in the know" about goings-on which the average "straight" heterosexual would probably not know about, nor want to know. Turner wove together no less than six major errors, or slanders, into that short paragraph, nearly all of which constitute dirty-minded smears:

1) He misrepresented the orgone accumulator's construction, as if it were just a simple "box".

2 & 3) He presented two falsehoods regarding the function of the orgone accumulator, repeating the old Brady-Gardner slanders of the 1940s, that it was promoted by Reich or his associates as something for a) "improving orgastic potency" or b) "improving mental health".

4) He misrepresented Reich's clinical scientific work and his orgone accumulator through the ideas and behavior of the "beat generation, sexual outlaw" cult novelists and poets Ginsberg, Kerouac and Burroughs, who made public distortions of Reich's work and ideas many years after his death. And just who are those guys?

A) Alan Ginsberg was a homosexual pedophile who wrote poems exalting his pathology. B) Jack Kerouac was a depressive bisexual alcoholic and drug-abusing mystic. C) William Burroughs was a bisexual alcoholic drug-dealer who one day casually shot his wife in the head with a pistol.

5) Turner's paragraph also misrepresented the Woody Allen "orgasmatron" device (a fictional electronic cabinet into which people went to obtain sexual excitement in a fictional future of widespread sexual abstinence) as something legitimately comparable to Reich's accumulator, which was in fact no "sexual appliance".

6) He suggested the malicious lie, that Neill or Reich, or both, might have locked up children in boxes. Taken in context with the sexual innuendo already within the paragraph, this also insinuates the sexual abuse of children!

This short 2004 *LRB* article set the tone and theme for Turner's later writings, in how he misrepresented Reich as some kind of twisted pervert and medical quack. The article was additionally curious for *LRB,* given how it was no "review" of any book – this was seven years before Turner's "Orgasmatron" book was published. Whatever the reasons, it indicated an early alliance against Reich between Turner and his editor at *LRB,* the very leftist Mary-Kay Wilmers, who openly boasts of Soviet spies and KGB assassins in her family tree.[3]

Turner wrote another smutty put-down of A.S. Neill and Summerhill School in 2010, in *Cabinet* magazine, for which he is co-editor. His various accusatory claims will be more fully addressed and refuted below, as they are repeated in grander style within his 2011 "Orgasmatron" book.

As a point of fact about which I will elaborate shortly, Reich considered bisexuality and homosexuality to be neuroses, and pedophilia a criminal act. He would spin in his grave to be compared with the likes of well-known but quite disturbed persons such as Ginsberg, Kerouac and Burroughs. Turner apparently selected them out for a false comparison to Reich, as a means to destroy him in the eyes of more rational, moderate and sexually healthy people.

Turner's 2011 "Orgasmatron" Book

Turner's 2011 book was titled *Adventures in the Orgasmatron: How the Sexual Revolution Came to America,* published by Farrar, Straus & Giroux of New York. A British edition appeared a few months later with the longer title *Adventures in the Orgasmatron: Wilhelm Reich and the Invention of Sex: How Renegade Europeans Conceived the American Sexual Revolution and Gave Birth to the Permissive Society,* published by Fourth Estate. These were followed by numerous prearranged "review" publicity articles in major mainstream print and internet newspapers and magazines, all of which repeated and glorified Turner's slander and defamation, happily expanding upon and amplifying them. The book capitalized upon the Woody Allen "orgasmatron" theme, being suitably sarcastic for skewering Reich, and sufficiently titillating to boost sales. While the subtitles suggest it is a work detailing social problems related to the excesses of the Western sexual revolution, in fact the book pushes the point that all of the pains and ills of Western sexuality can be blamed upon one person: *Wilhelm Reich.*

The book is given a false aura of dignity via self-congratulatory claims by the publisher, that it is an "exhaustive biography" or even "the first authoritative biography" of Reich, which is *completely false*, a marketing ploy only. The claim ignores several authentic historical biographies and scientific texts on the subject of Wilhelm Reich's life and work, such as Myron Sharaf's *Fury on Earth: A Biography of Wilhelm Reich*, or David Boadella's *Wilhelm Reich, the Evolution of His Work.*

With *Orgasmatron* in the title, and a cartoon image of a woman's

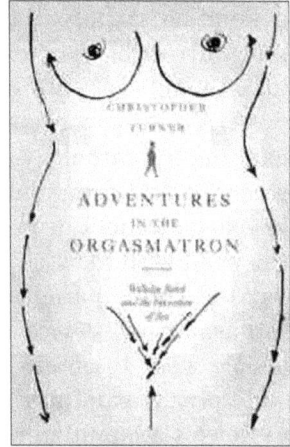

naked torso on the covers, the serious reader is warned this is not anything of scholarly merit. It is similar to how romance novels always have a conventionally attractive couple on the cover, locked in a half-naked kiss. The porno magazines exploit this method more exactly, and it is hard to say if the Turner book could do better. You won't miss it on the nonfiction shelves, though it should have been placed into the modern fiction or satire category.

But the title deceives in other ways, as this is no lighthearted humor at work. It pushes the reader towards a direct but fully false identity between Reich's orgone energy accumulator – a device for health maintenance, and scientific biological or physical research – and the fictional "orgasmatron". The book repeatedly tries to hammer home that fiction: that Reich's work was purportedly focused upon some kind of fantastic gadgetry aimed at "orgasm-energy", by which he could be casually dismissed as a "crackpot, pervert and fraud", or even "mentally ill", with no legitimate scientific findings worth a damn. Turner's narrative is a complete sham, however, as Reich was none of those things.

Turner's Background and Credentials For this Effort

What are Mr. Turner's background and credentials, to take on such a task as the full-bore analysis of the character and life-work of Reich? His biography is summarized online by his publicist,[4] who informs us there's nothing of natural science or historical research in Turner's background. He's a student of the literary arts, with a year 2000 media-arts certificate from the *London Consortium*, his graduate thesis being focused upon "the cultural history of disgust" – this is no joke, the exact title being *The Disgusting: The Unrepresentable From Kant to Kristeva*.[5] This literary achievement was appropriately followed by consulting work in London theater. His published works include a collection of book reviews, a newly acquired editorship of *Modern Painters* magazine, and he posts up his own magazine-blog *Cabinet*, which inflates the term "literary intellectual" to new heights. The Summer 2011 issue of *Cabinet,* for example, carries articles by his associates on such unserious topics as *"Key Dates in Clown Crime"*, *"People's Friendship Salad and Other Culinary Expressions of Brotherhood"*, *"Drab Colors"*, *"The Ethics of Rubbernecking"*, *"No, Oedipus Does Not Exist"* (a bit of denial?), and another by Turner skewering his father-figure Reich.

In Defense of Wilhelm Reich

Since publishing the 532-page "Orgasmatron", he's been rewarded with additional publishing opportunities at *The Guardian, Sunday Telegraph*, and of course the *LRB*, all of which are decidedly leftist in slant, a common characteristic for Reich's major detractors.

With that foundation, or lack thereof, Turner undertook a multi-year odyssey to investigate Reich – so it is claimed – but the finished product suggests this was mainly a search for dirt. As I will show, Turner treats the subject of his study – Wilhelm Reich – with pure contempt, mixing up a few facts with salacious innuendo and slander, never giving him the benefit of the doubt on poisonous but scantily documented, or fully undocumented accusations. What emerges is a completely false history.

Let's examine Turner's major claims and charges against Reich in the "Orgasmatron" book, starting with the concentrated poison found within his *"Introduction"*. He begins his discussion with a series of falsehoods which build upon each other towards a crescendo at the end of the book. As is typical when a dozen false claims or slanders are woven into a few short sentences, it will require considerably more space to substantively refute them.

Repeating Old Slanders from Brady and Gardner

CLAIMS (p.4-5):§ *"...Reich...arrived in New York in late August 1939...optimistic that his ideas about fusing sex and politics would be better received there than they had been in fascist Europe. ... Reich was a sexual evangelist who held... the orgasm was a panacea to cure all ills...[and] would lead to a 'genital utopia'..."*

FACTS: Reich began to move away from his organized public SexPol reform efforts in Europe starting around 1933, a process that was essentially completed by the time of his 1939 departure to the USA. He was then focused upon new experimental findings on the bioelectrical functions of life,[6] on the microscopical bions,[7] and on his newly developed ideas on the orgone energy radiation and accumulator.[8] Even during his SexPol years, Reich was never a "sexual evangelist" as Turner invents. He in fact had become deeply unsettled about the great difficulties of sexual reform work, and how quickly it had all been swept away following the rise of Nazi

---

§ All unidentified page numbers reference the American edition of Turner's 2011 "Orgasmatron" book.

fascism, and the betrayals of freedom movements by the European Communist Parties. Around the time of his arrival in America, his work ethic is best understood by what appeared in his new journals, or on signs hung on the walls of his new laboratory, stating: *"No Propaganda is the Best Propaganda"*, *"Work Not Politics"* and a more optimistic remark about his scientific approach, *"It Can Be Done"*. He focused upon laboratory research, training therapists in his new invigorating emotion-release methods, and he wrote and published, but did not try to build up a social movement as he once had attempted in Austria and Germany.

Turner correctly points out that Reich's contemporary critics contemptuously mocked him, if not in Turner's exact words about his seeking a "genital utopia" or "orgasm panacea" – but these anxiety-laden put-downs of orgastic functioning never accurately described Reich's findings or goals, as Turner implies. Such statements by Reich's critics were merely the usual jackass brayings and barkings that Reich witnessed in those years, for daring to openly and seriously discuss sexual issues, or for having "sex" or "orgasm" in the titles of his publications. The fact that Turner can today publicly use the word "Orgasmatron" for his book title, even with a naked woman cartoon on the cover, without risking his own neck or job, is a testament to how the old obstructionist prudery of past decades has been swept away. Female orgasm was hardly considered existent before Reich wrote positively about it in the 1930s. As Reich noted, concepts of male sexual potency were then (and still are) confusedly tied up with mere erection or ejaculation, and boasted about in popular phallic narcissism. Reich was attacked for daring to write openly on such subjects, exposing people's various fearful, neurotic or posturing braggart attitudes about sexuality. And as is typical in such situations, his critics reversed the entire meaning of his work. So Reich, who was against pornography and the "free fuck", is mentioned by Turner alongside Henry Miller's declaration about Europe being "the Land of Fuck," (p.5) deceptively implying Reich would have approved.

CLAIMS (p.5): Speaking about the orgone accumulator, Turner describes it as *"...a wooden cupboard about the size of a telephone booth, lined with metal and insulated with steel wool... Reich considered his orgone energy accumulator an almost magical device that could improve its users' 'orgastic potency'... [which] could not only help dissolve repressions but also treat cancer, radiation sickness, and a host of minor ailments."*

FACTS: Here, Turner's inaccurate description of the orgone accumulator comes straight out of the malicious 1940s and 1950s writings of pestilent leftist writers Mildred Brady[9] and Martin Gardner,[10] mixed with his own earlier venom from 2004. Several false statements are all mixed up, including the false claim that Reich considered the accumulator to be a "magical device" able to treat just about anything, including sexual impotence and cancer. While he developed an *experimental cancer therapy*,[11] and showed it could enhance one's *resistance to disease* (today that term is largely replaced by *immunity*), he never claimed the accumulator was such a "cure-all".

As discussed in prior chapters, the orgone accumulator resembles a Faraday cage,[12] and wood is used in the supporting structural framework but not as a primary functioning component. It is more authentically composed of alternating layers of *ferromagnetic metals interlain with dielectric insulating material* such as sheep's wool or fiberglass. Dielectric materials are typically alternatively interlain in the walls with sheets of steel wool. A final layer of sheet metal steel or galvanized steel is added to the interior, with a final exterior layer of sturdy insulating dielectric. This type of structure also resembles the layered nature of an *electrolytic capacitor*, except it is much larger and has a hollowed-out space within the interior, with a door to enter or exit. As such, it has the additional properties of what I call a *hollow capacitor,*[11] able to spontaneously build up a higher charge density within the interior of the accumulator. This higher charge was documented by Reich through its biological effects and physical properties, as discussed in prior chapters. His findings were documented by objective clinical applications, laboratory experiments, and other empirical methods.

The orgone accumulator was used *experimentally* for a variety of health problems, mostly related to weakened disease resistance and a low energy level, but *Reich never claimed it could "cure" cancer or sexual impotence or emotional repressions which, as he also discovered, had much deeper roots in the person's past biophysical history and character structure.* These facts are not clarified by Turner, whose distortions make Reich out to be a purveyor of fantastic claims.

Turner certainly knows this history of unethical smears and lying against Reich, so it beggars belief that his repetition of the worst of the malicious 20th Century misinformation might be accidental. Here are a few quotes from Reich's publications, written in response to the 1947 Brady smears:

> "The orgone accumulator, as has been clearly stated in the relevant publications . . . cannot provide orgastic potency." [13]

> "In Collier's, May 26, 1951, pp. 53-54, the erroneous statement is made that Reich claims to have a cancer cure, a slander originated some years ago by Mildred Edie Brady in The New Republic. Reich does not now nor did he ever promise any cancer cures. Reich's research on the cancer problem, including his solution of the riddle of the origin of the cancer cell, is described in The Discovery of the Orgone, Vol. 2: The Cancer Biopathy." [14]

> "In 1946 several Russian sympathizers gathered in the assault upon orgonomy. Wertham, a physician, who was a member of the American Russian Friendship Association, a clear cut 'front organization', published his attack in The New Republic, a left wing, then Russian-minded periodical. Mildred Brady then took over by planting the slander of 'orgastic potency' having been promised to a gullible public, an accusation easily disproved. Brady was followed by the FDA, an agency which had shown no interest in the discovery whatever until the Red Fascist sympathizer had reared his ugly head.
> ... the old Mildred Brady slander, namely the alleged function of the accumulator [was] to provide 'orgastic potency.'
> ... distortion and slander, contrary to truth. No one had ever promised to cure the widespread disease of 'orgastic impotence' by way of the accumulator. [8]
> ...The pornographic slander started by Brady, that the orgone energy accumulator 'conveyed orgastic potency' was now connected with 'fraud' and with the Brady slander of 'cancer cure.'" [15]

Today, Turner willfully repeats the same old slanders. It is important to note that Reich was 100% correct about the Soviet sympathies, and even the cloaked Communist Party backgrounds of his American detractors Wertham[16] and Brady.[9] Both published their attacks in *The New Republic* magazine, which at the time was run by Michael Straight, a self-confessed American member of the notorious *Cambridge Soviet spy ring*. The magazine was then edited by Henry Wallace, a Soviet mole who had ascended to the top position of Vice President of the United States under President Roosevelt, until his pro-Soviet declarations led to his being replaced by Harry Truman. Mildred Brady and her husband Robert were both fired years earlier from the Roosevelt *Office of Price Administration* for their open Soviet sympathies. She had been subjected to FBI surveillance which revealed connections between her and her husband Robert Brady, to the KGB-run Silvermaster spy ring, which took atomic secrets to the Soviet Union. All this and much more is documented in the book *Wilhelm Reich and the Cold War* by James Martin,[17] who used Freedom-of-Information-Act searches of FBI and other government agencies on Reich's detractors. Martin began his investigations believing he might refute Reich's assertions about a communist conspiracy, but instead wound up confirming them rather exactly.

Reich's detractors have overwhelmingly been *Reds*, and some of the same was at work within the FDA as well. I've already provided more information on this subject in the previous Chapters. Reich had escaped the Nazi and Soviet death machines in Europe, only to be attacked years later by Red operatives working clandestinely in the USA, under the cloak of "consumer protection", "progressive liberalism", "educating the public", and "fighting medical quackery".

Turner and his major editors and facilitators, it should be noted, also appear to come from similar hard-left backgrounds, pushing for Big Government controls over the lives of ordinary people, to "regulate them down into their underwear" as was a popular saying by ordinary people living under the Communist regime of Cold War East Germany. Today, however, they cloak or self-deceive as merely "liberal", which is decidedly *not the same thing*. Authentic *liberals* respect truth and facts, just as do *conservatives*. They also both oppose defamatory slander, public lying and book-burning, even if done to support causes they agree with. *Leftists* by contrast, are similar to *extreme-right Nazi types*, in that *they have no problems with lying, censoring the truth, or maliciously attacking anyone who stands in their path to power*. They also have no hesitation to burn books and imprison, or kill those expressing dissenting opinions.

After slandering Reich by claiming he said the orgone accumulator was *"...an almost magical device that could improve its users' 'orgastic potency'"*, Turner corrects himself in a citation, where he writes that orgastic potency was only the *end-goal* of Reich's emotion-release therapy methods (citation 7, p. 5, 450). This indicates Turner knows the orgone accumulator was never offered for such "orgastic" purposes. Since he knows this, but hideously distorts it in the text-body of the book, it stands out as a deliberate deceit. To the casual reader who might not bother to flip from page 5 to page 450 and review the citation, it appears as if Turner's ugly text description is accurate merely because it carries "a citation", and because Reich's *Function of the Orgasm* [18] is mentioned on the same page! But there is *nothing* in Reich's work, *anywhere*, which supports Turner's malicious repetition of Brady's 1947 slanders.[9] Reich never said such things as Turner claims, about the orgone accumulator construction, nor its functions, nor experimental use.

Here, for example, is Reich's own description of *orgastic potency*, taken from the Glossary of his *Function of the Orgasm*, as reproduced from the 1947 edition,[18] which is also repeated in later editions published after his death.

> *"Orgastic Potency. Essentially, the capacity for complete surrender to the <u>involuntary</u> contractions of the organism and complete discharge of sexual excitation in the acme of the sexual act. It is always lacking in neurotic individuals. It presupposes the presence or establishment of the genital character, i.e., the absence of a pathological character armor and muscular armor. The concept is essentially unknown and usually not distinguished from erective potency and ejaculative potency, both of which are nothing but prerequisites of orgastic potency."* [18]

Reich and Einstein, Misrepresented

CLAIMS (p.6): *"Reich persuaded Albert Einstein to investigate the [orgone accumulator] machine, whose workings seemed to contradict all known principles of physics, but after two weeks of tests Einstein refuted Reich's claims".*

In Defense of Wilhelm Reich

FACTS: Reich did meet with Einstein in early 1941. He eventually published their extensive letter exchange in a document titled *The Einstein Affair*,[19] which is the primary source on this matter. The two men originally met on 13 January for a five-hour meeting and discussion on Reich's discoveries, by Einstein's invitation to his home in Princeton. Einstein was so intrigued with Reich's demonstration of a small orgone accumulator and orgonoscope (which allowed direct visual observation of the orgone energy) that he asked to borrow them for a period. Reich consented. Einstein then ran some independent experiments, and three weeks later on 7 February, he wrote back to Reich:

> *"I have now investigated your apparatus, but I have limited myself to the temperature phenomenon, because I was not able to exclude subjective impressions with regard to the light manifestations. In the beginning, I made enough readings without any changes in your arrangements. The box thermometer showed regularly a temperature of about 0.3 - 0.40 higher than the one suspended freely."* [19]

Einstein therefore *confirmed* the thermal anomaly within the orgone energy accumulator. He also confirmed, though without clarity, the subjective light phenomenon inside the accumulator. He had stated, during their original meeting, that Reich's findings would be "a great bomb" for science, so he knew their controversial nature. However, while he was man enough to admit Reich was correct about the experimental anomaly, he too-easily accepted the arguments of an unnamed assistant, that the thermal anomaly was created by convection currents in the room, and not by an orgone energy effect.

Einstein's letter also revealed that he or the assistant partly disassembled the accumulator apparatus Reich had left with him, thereby *rendering it inoperative* and, not unexpectedly by orgone theory, yielded no results. In that same letter, Einstein wrote:

> *"I started by taking the metal box out of the enclosure with the windows, put it directly on the table and fixed the thermometer on the top side of the box by putting it in the now somewhat larger hole in which the earthenware pipe was before, using a folded newspaper to make the thermometer stick. Result:*

(gained by many experiences) The temperature difference disappeared or was, in any case, not more than + /- 0.1 °." [19]

The "metal box" Einstein references would be the internal metal layer of the apparatus, which by itself does not constitute an orgone accumulating structure. An orgone accumulator would require an external organic-insulating exterior to function, just as an electrical capacitor requires both its conductive and insulating components to function. With the outer dielectric layer removed, Einstein or the assistant were forced to use a scrap of paper stuffed around the thermometer receptacle hole to make it fit, which suggests the precision of Reich's instrument was further degraded. Einstein's confused effort to understand the thermal anomaly via disassembly of the accumulator was a critical error which wiped out the phenomenon he had earlier measured and confirmed.

Reich patiently replied to Einstein on 20 February,[19] with a letter giving detailed instructions on control experiments to rule out such convection effects. He also clarified the necessity of the external cover for the experiment to function properly. However, Einstein never responded to this letter, and appears to have never undertaken those critical control experiments. The reason for Einstein's subsequent silence remains a mystery.

Einstein may not have undertaken the necessary control experiments on the orgone accumulator thermal anomaly experiment, but many others have, with positive results overall. My own multi-year replication of Reich's thermal experiment, using modern measuring apparatus, fully ruled out convection effects by use of a control enclosure thermodynamically matched to the orgone accumulator. I also placed both accumulator and control on a slowly rotating platform under a fully shaded measuring shelter. My experiment allowed around-the-clock data acquisition using thermistors, calibrated down one thousandth of a degree. A pulsating thermal anomaly of approximate 0.5°C. was clearly measured and documented over several years of the experiment, with occasional peaks of up to 0.7°C. My published paper on this subject cited six other scientists who had made similar positive replications.[21]

Was Turner so ignorant of all of this? Or did he just willfully and unethically decide not to mention any of it?

Turner includes Reich's document *The Einstein Affair*[19] in his bibliography, so assuming he actually read it, he would know about these issues. He also cites papers from the *Journal of Orgonomy*[20] where it suits him. Other studies, including experimental reports on my positive multi-year replication of Reich's thermal anomaly experiment, were published and available online as early as 2009.[21] Even a ten-minute internet search would have brought up details on this subject, so his failure to accurately recount what Einstein actually wrote, or to mention Reich's important rebuttal, or to mention other positive replications of the thermal anomaly experiment, appear as deliberate *Lies of Omission*.

Influencing America: Reich versus Kinsey

CLAIMS (p.6): Turner adds several false statements immediately after the one on Einstein, on a theme which recurs in his book:

"...the orgone box became fashionable in America in the 1940s and 1950s, when Reich rose to fame as the leader of the new sexual movement that seemed to be sweeping the country."

FACTS: Reich was *never* the leader of any "new sexual movement... sweeping the country." Nor did he "rise to fame" in America, nor did the orgone accumulator ever become "fashionable". Turner inaccurately elevated Reich into a position which more factually belongs to Alfred Kinsey, a man whom Turner admires greatly and writes about at length in other publications, but who is down-played in the "Orgasmatron" book for reasons which I believe are central to his motivations in attacking and trying to destroy Wilhelm Reich.

Kinsey was the man of the mid-20th Century getting tremendous publicity for his sexual theories and shaky claims about American sexual behavior, as printed in his explosive 1948 book *Sexual Behavior in the Human Male (SBHM)*.[22] Kinsey was heralded in public as the "father of the American sexual revolution" and his work compared with the atom bomb. He was constantly giving university lectures, magazine interviews, and was consulted by government agencies. His ideas and theories were used to redraft laws regarding sexual behavior. According to the Kinsey Institute website, *SBHM* sold over 200,000 copies in the first year of publication, and went on to become a *"bestseller... an international sensation"*.[23] His later *Sexual Behavior in the Human Female (SBHF)*[24] added to his fame and fortune. They were celebrated by the mainstream press as being among the most influential of scientific books for the 20th Century.

By contrast, after Reich gave up his teaching position at the *New School for Social Research* and built his *Orgone Institute* near to Rangeley Maine, he lived in relative isolation. While Reich was a prolific writer, he was forced to self-publish, was not invited to speak at universities, was never consulted by government agencies, and only rarely did others seek him out for interviews. Aside from the smear articles which erupted after 1947, he remained in relative obscurity, working productively and quietly. His books sold only modestly as compared to Kinsey's books, although many professionals did help to extend the influence of Reich's ideas through their own private practices.

Reich's Orgone Institute issued a report in 1952 indicating that, over the prior six years, approximately 35,000 copies of all of Reich's books and journals combined had sold.[25] That means around 5,800 copies were sold annually. That's not so bad, and we know Reich's books have been slow steady sellers over the years. But it hardly approaches the attention given to Kinsey's books during that period. At the height of Reich's working life, there were only about 300 orgone accumulators being rented out to patients by his nonprofit institute, as a means to fund his research, and then only by prescription from the medical doctors working under his direction.

Reich's ideas and works never became anything so "fashionable" as those of Kinsey, by any account, and the world wasn't exactly beating a path to his door. Due to hostility within the mainstream media and medicine of his day, only a very small percentage of the American population even knew about Reich, except for the salacious media smears after 1947, which did spread widely and distorted everything.

As previously noted, it is also false to say that Reich was trying to build a "sexual movement" in America. He found American social conditions far better than what he had left behind in Old Europe, and was dedicated to his new scientific and experimental work. By contrast, Kinsey relished the limelight and hob-knobbed with all the high-up fashionable people in Hollywood, Big Media and Washington DC. Kinsey got all the public laurels.

So why would Turner wrongly assert that Reich was a major influence in American society on sexual matters, while ignoring Kinsey and other factors? The subtitle of the American edition of his "Orgasmatron" book was *"How the Sexual Revolution Came To*

In Defense of Wilhelm Reich

America". The subtitle to the British edition by Fourth Estate was more explicit: *"How Renegade Europeans Conceived the American Sexual Revolution and Gave Birth to the Permissive Society"*. Clearly, Turner and his editors view Reich as a "Renegade European" who "Gave Birth to the Permissive Society". Kinsey wasn't a European, and the book hardly mentions any other Europeans besides Reich, except to reference their slanders of Reich. There is no factual basis to give Reich such a highly influential position. So why make such a claim?

The answer may be found in what the Turner "Orgasmatron" book *omits* about Kinsey and the *very real pornographic sexual degeneracy which frequently appeared on the American social scene as a consequence of the Kinsey agendas*. To address this issue, I must provide several pages detailing Kinsey's work, and the lesser-known but dramatic differences between Reich's and Kinsey's activities, theories and persona.

Turner's Serious Omissions On Kinsey

Turner's 532-page book postures to detail the American sexual revolution, but devotes only 15 pages to Kinsey. In every case, Kinsey is presented in a positive light, but without any elaboration about just what Kinsey was trying to accomplish, or his theories on sexuality. No clarifying details are given. This is strange indeed, for the singular fact that Kinsey has for a long time come under serious public criticism for having ushered in an era of extreme pornography and unfettered sexual license on the social scene. Kinsey's efforts did not merely aim to decriminalize neurotic sexual expressions between consenting adults, such as homosexuality and bisexuality, but to "normalize" them, as well as other severe pathological sexual impulses such as sadomasochism, bestiality and pedophilia.

This darker side of Kinsey was first exposed by scholar Judith Reisman in her book,[26] *Kinsey, Sex and Fraud* (1990) and by Reisman and therapist Edward Eichel in *Kinsey: Crimes and Consequences* (2004). British Yorkshire TV produced a special 1-hour program on the subject in 1998, titled *Kinsey's Paedophiles*.[27] This documentary carried confirming interviews with some of those who worked with Kinsey, as well as with the now-grown children who were subjected to sexual abuse by pedophiles enrolled in

Kinsey's data-collection schemes. The *Kinsey Institute* has engaged in what appears to be a full-tilt denial and cover-up of these facts, aided by a generally Kinsey-sympathetic (and porn-sympathetic) mainstream-left media. Kinsey's official biographer, James H. Jones, however, confirmed these staggering facts in his 1997 work *Alfred C. Kinsey: A Public/Private Life*, as they could no longer be denied or kept secret.[28] Kinsey was on record, repeatedly, in defending pedophiles and advocating decriminalization of their rapine practices. He offered little sympathy to the child victims of such horrific abuse.

In the "Orgasmatron" book, Turner heaps praise upon *"Kinsey's renowned investigations"* (p.4) and *"his monumental study of American sexuality, which would document in enormous and controversial detail a country with licentiousness seething below its prim facade."*(p.210)

But in fact, Kinsey's "renowned investigations" and large claims never stood up to serious scientific review. His data as first published in *SBHM* proved to be deeply flawed, notably in his declarations that 8% to 54% of the American male population was actively, or at some time in the past had been engaged in homosexual activity.[29] That was never supported by other investigators, who have put the number closer to something less than 2%.[30] Kinsey's other eye-popping claims, such as 11% of rural young men working on farms had engaged in bestiality,[31] or that the American population was flooded with concealed armies of prostitutes servicing high percentages of the male population, could never be supported or verified by others. The figures could be *understood,* however, as the product of biased data collection, shoddy statistical work, and Kinsey's vivid imagination.

As we know today, Kinsey's interview subjects were seriously oversampled from sex-offenders in prisons, prostitutes and pimps, men in homosexual bars and nightclubs, and even more seriously (and secretively), from *active criminal pedophiles and other sexually-disturbed individuals.*[26,27,28] This was aside from the issue on how people of more modest and healthy, natural heterosexual behavior (Turner's so-called "seething prims") were not likely to be in prison for sexual crimes, nor hang out at homosexual bars or seedy hotels servicing prostitutes, where Kinsey did a significant part of his data collection. Nor were they likely to consent in the first place to probing interviews with a stranger on their private sexual lives. A certain quality of exhibitionism or boasting sexual narcissism was required for such too-easy consent to be interviewed. Consequently, Kinsey's data and conclusions reflected only the behavior within those select groups he sampled from. As the computer technicians would say: *garbage in, garbage out.*

Kinsey's preference for seeking out interviews with those engaged in kinky or criminal sexual conduct, rather than attempting to seek out random samples from the general public, reflected a darker aspect of his personal behavior. In public, Kinsey and his wife presented a stable married heterosexual couple, as normal as apple pie. To the outside world, he was a serious professor at Indiana University, whose Institute was funded by the Rockefeller Foundation. Privately, however, he was a deeply troubled bisexual who preferred homosexual liaisons. He stacked his Institute with associates of similar sexual preferences, and privately developed secret working associations with serial pedophile child-rapists, even as they continued to commit their crimes. The most serious charges against Kinsey emerge in this latter respect.

Kinsey's "data" on childhood sexuality was largely derived from written or oral reports given to him by child molesters and rapists with whom he had corresponded, and had encouraged to provide him with increasingly elaborate details. His pedophile contacts did not disappoint, and with Kinsey's encouragement – including instructions from him on how to "collect data" – they went on to sexually molest and rape hundreds of little boys and girls. These "data" were then neatly assembled in Tables 31-34 in his *SBHM,* for child-rape victims ranging in age from a few months to 14-years.[32] The data are extremely detailed, identifying such things as "orgasm in 3.5 minutes" or "multiple orgasms" over a 24 hour period in some cases, indicating a days-long session of child-rape. Kinsey dutifully wrote what the pedophiles had reported to him, including how the child victims' reactions to their "orgasms" included *"violent convulsions of the whole body; heavy breathing, groaning, sobbing, or more violent cries, sometimes with an abundance of tears (especially among the younger children)."*[33]

Kinsey's book on *Sexual Behavior in the Human Female* (*SBHF*) carried similar indications that girl children were also sexually abused, as seen in Tables 25, 26, 31, 32 and 35 of *SBHF*, detailing masturbation by adult pedophile note-takers of girls as young as 3 to 5 years old.[24]

For Kinsey, this monstrous picture of violent old men raping hysterically screaming and struggling little children was, however, only "data", and his emotional sympathies clearly went out to the pedophiles, as "misunderstood victims of sex-repressive society". Little concern was expressed for the terrified and destroyed children.

These materials are hardly known today, however, given how the liberal-left mainstream media (the same media which chronically skewers Reich and sneers at "Christian prudes") continues to glorify Kinsey. In 2004, a major Hollywood film about Kinsey was produced, typically paving over these and related issues.

Kinsey's ideas and claims were also abused for mutual self-promotion by the Hefner *Playboy* group and more extreme porn magazines. They incrementally pushed the limits way beyond mere "naked girl" magazines, into public endorsement of homoeroticism, bisexuality, sadomasochism, bondage, pedophilia and child-porn, all of which have exploded via internet. Pornography today, both in print and online, contains gang-rape images, including "fantasy" or on the black market, fully real imagery where kidnapped women and children are violently beaten and sexually violated, or worse, in front of video-cameras. San Francisco and other cities boast of sometimes barely-legal internet studios specializing in such material, servicing sexually-disturbed psychopathic audiences worldwide.

Turner is no innocent "babe in the woods" supposedly ignorant about such things. His writings almost specialize in exotic sexual material, and where he writes on the subject of Kinsey, it is only to glorify or defend him, but never to expose to his readers just what are the critical charges against Kinsey. In 2010 Turner wrote a basically friendly review of Hugh Hefner's £900 six-volume autobiography, noting how *"Hefner says it was the publication of Alfred Kinsey's sexual surveys that provided the impetus for Playboy.*[34] Hefner later funded some of Kinsey's research. But as related above, Kinsey provided legitimacy for a whole lot more than mere pictures of the naked body.

Turner's admiration for Kinsey appears uncontainable. In the "Orgasmatron" book he reports almost joyously Kinsey's interviews with men from the hyper-promiscuous homosexual nightclub subculture, where some boasted to have *"...enjoyed as many as two thousand to three thousand partners each."* (p.212) Turner describes Kinsey's interviews in London brothels and with European transvestites as something absolutely heroic, and evidence for an

Kinsey's Notorious *Table 34*

Table 34 is composed of "data" collected under his supervision by pedophile child molesters and rapists, who meticulously wrote everything down in notebooks, presenting them to Kinsey for use in his famous *Sexual Behavior in the Human Male*.[24,32] Note the age groups range from infants to 14-year old boys, who were dutifully recorded as struggling, screaming and crying hysterically[33] while being subjected to physical sexual assault by Kinsey's pedophiles, stopwatch in hand, for up to a full 24 hours. Kinsey's "data" on "child orgasms" and "multiple orgasms" was thereby obtained. This unscientific and criminal calculus forms the foundation of Kinsey's toxic views on childhood sexuality and the "normalcy" of adult-child sexual activity. Similar problems trouble his entire body of work.

AGE	NO. OF ORGASMS	TIME INVOLVED	AGE	NO. OF ORGASMS	TIME INVOLVED
5 mon.	3	?	11 yr.	11	1 hr.
11 mon.	10	1 hr.	11 yr.	19	1 hr.
11 mon.	14	38 min.	12 yr.	7	3 hr.
2 yr.	{ 7 / 11	9 min. / 65 min.	12 yr.	{ 3 / 9	3 min. / 2 hr.
2½yr.	4	2 min.	12 yr.	12	2 hr.
4 yr.	6	5 min.	12 yr.	15	1 hr.
4 yr.	17	10 hr.	13 yr.	7	24 min.
4 yr.	26	24 hr.	13 yr.	8	2½ hr.
7 yr.	7	3 hr.	13 yr.	9	8 hr.
8 yr.	8	2 hr.	13 yr.	{ 3 / 11 / 26	70 sec. / 8 hr. / 24 hr.
9 yr.	7	68 min.			
10 yr.	9	52 min.	14 yr.	11	4 hr.
10 yr.	14	24 hr.			

Table 34. Examples of multiple orgasm in pre-adolescent males

Kinsey's views clashed with those of Reich, who demanded children be protected from seducers and rapists. Kinsey also disagreed with Reich about the primacy of full genital heterosexual orgasm, declaring quite openly that *any kind of ejaculation by a male constituted "orgasm"*, even if premature, anger-laden or unpleasurable – and all Kinsey "orgasms" were of equal physiological and psychic value no matter how they were achieved. Author Turner celebrates Kinsey's "research", even while hypocritically slandering Reich with smut in the "Orgasmatron" book.

"enlightened sexual paradise." (p.359)   In relating one of Kinsey's research trips to New York City, Turner delightfully writes: *"The house detectives at the Astor Hotel in New York objected to the prostitutes, pimps and drug addicts who were piling into the lobby from Times Square to catch the elevator up to Kinsey's room so as to earn the dollar he would give them for recounting their sexual histories."* (p.264) Turner uncritically recounts Kinsey's meetings at that time with Burroughs, Ginsberg, Kerouac, and a male prostitute they were hanging out with, and how they abused the orgone accumulator for sexual purposes – but without any sense of the pathology in their conduct. (p.263-264)

All these Kinsey anecdotes are abused by Turner as various lead-ins to denigrate Reich, implying or declaring that Reich would have approved, as where Turner describes Kinsey's New York trip and the *"libertine circle, as they [Burroughs, Ginsberg, Kerouac] dubbed themselves, [addicted] to drugs, sailors, porn, bathhouses, and Wilhelm Reich"* (p.263) – one might forget, Reich never knew any of these characters, and was critical of such sexually neurotic and self-destructive lifestyles. But Kinsey did meet with them, and approved. And apparently, so does Turner.

As previously noted, Turner is an editor of *Cabinet* magazine, which carries a series of articles along these lines. *Cabinet* #8 of Fall 2002 provides a friendly book review of *Welcome to the Molly-House* by Bailey and Trumbach.[35] For those who don't know, a "Molly House" is a place of homosexual prostitution or an early version of the "hookup bar", where drinks are sold and homosexual liaisons – frequently with boy prostitutes – take place in back rooms. The article nonchalantly speaks about men picking up "teenage boys" or "young boys" for pedophile sex, in the "chapels", as they call the back-rooms. This is no critical article' it carries the suggestion of an endorsement of the "normalcy" of pedophilia. The only implied complaint appears to be that it has to take place hidden, "underground".   I could not find any distancing "editors note" by Turner or any other *Cabinet* editor, indicating that they might disapprove.

Turner's most personally damning discussion of Kinsey, however, appears in a 2005 *London Review of Books* (*LRB*)[36] article. There, he combines a review of the Hollywood film about Kinsey as directed by William Condon (actor Liam Neeson playing Kinsey) with a defense of Kinsey against his critics, including the most serious accusations of Kinsey encouraging and gathering data from active pedophiles. In defending Kinsey, Turner's methodology is limited to snide commentary, selective recounting of history, and attacks on the motives of the primary researcher and critic who originally exposed Kinsey's actions to the light of day, Judith Reisman.[26] With backhanded dismissals, false exaggerations and personal attacks, he tries to make Reisman's charges sound unbelievable, but without addressing their substance. He derisively writes:

> *"Reagan's porn tsar...she equates Kinsey with Josef Mengele, accusing him of skewing his statistics to legitimise all sorts of perversions, of kidnapping and drugging young boys in order to conduct secret sexual experiments on them, of soliciting paedophiles to record their exploits for him, and of making and collecting child pornography... Reisman sought to preserve the purity of childhood by recasting all Kinsey's writing about childhood sexuality as abuse."*[36]

In fact, Reisman did not merely "accuse" but *factually documented* her charges against Kinsey, from facts taken straight out of his own books, as I've confirmed and presented here. Turner rather blandly recounts some of that sordid history, as with Kinsey's relations to "Mr. X", a pedophile who had confessed to raping 600 preadolescent boys and 200 preadolescent girls. Those sadistic criminal actions were all dutifully reported to Kinsey in personal interviews and in a notebook.

By contrast, the Kinsey Hollywood film depicts Wardell Pomeroy, one of Kinsey's assistants, reacting to "Mr. X" with outrage, and Kinsey angrily denouncing pedophilia – but those scenes are a Hollywood revision of history, a fabrication to rescue Kinsey from his true opinions and actions. However, Turner's LRB article expresses no outrage at all, but merely minimally recounts "Mr. X's" crimes in a curious, matter-of-fact manner, as if the reader should find them fascinating.

Turner then writes that the film's Pomeroy was merely *"voicing Reisman-like disgust"*, as if it was an impolite thing to express outrage about a pedophile who raped hundreds of preadolescent children! In fact, the real Wardell Pomeroy never expressed any such disgust, so far as I can determine, suggesting "Mr. X" was of great fascination for the entire Kinsey crew. Turner condemns

Director Condon for inserting the invented scene of "Reisman-like disgust" into the Kinsey film, declaring it to be an "...*attempt to give his story a moral backbone...*" ... as if moral backbone against serial pedophiles raping hundreds of children was a bad thing.

And that's the problem with Turner's review of Kinsey's life and work, and his review of the film: he never once expresses any serious concern or outrage about such brutal rapine and criminal activities, nor empathy for the child victims. In the "Orgasmatron" book, just as the facts about Reich's clinical and experimental work are erased from history *to better condemn him*, so too are the facts about Kinsey's pedophiles erased *to better uplift him.*

Turner grants Kinsey clean credibility for conduct inexcusable for any scientist (or anyone else), even while dismissing Kinsey's critics with open contempt. This is an early example of his slippery evasiveness, reflecting what he later does to Reich in the "Orgasmatron" book, except that Reich committed no such crimes and was explicitly against the sexual pathology that Kinsey promoted, facilitated, and normalized.

There's more along this same line, of the sheer hypocrisy of Turner and friends delightfully and uncritically reporting on pornographic promiscuity by Kinsey, Burroughs, Ginsberg, etc., on the one hand, while on the other hand slandering and falsely condemning Reich for alleged pornographic promiscuity he never engaged in nor endorsed.

On 27 June 2011, for example, the online *Slate.com* magazine ran several articles which exposed this hypocrisy exactly. One of the articles was a book review of Turner's "Orgasmatron", written by Peter D. Kramer,[37] a psychiatrist who has advocated for a strong genetic component to both depression and homosexuality. Not surprisingly, Kramer is also a supporter of drug therapy for emotional depression. He wrote *Listening to Prozac*,[38] endorsing a risky drug notoriously used to put a smiley-face on people's emotional misery, and which is wildly celebrated among pharmaceutical company stockholders. Such legal psychopharmaceuticals are especially widely prescribed for women with low self-esteem and trapped in loveless marriages, to suppress their sexual misery and avoid the messy business of divorce, or otherwise removing themselves from difficult emotional traps. They are also drugs of choice for the many "gay" men suffering from similar emotional traps and sexual misery. Naturally, with that background, Kramer has no love for Wilhelm

In Defense of Wilhelm Reich

Reich, whose work on the *functional relationship between psyche and soma* provides a direct pathway by which both biochemistry and neurology are under the control of emotional bioenergetic and social factors.[6,18] In any case, Kramer found Turner's book to be a useful tool for attacking and discarding Reich, in yet another false narrative of his claimed degeneracy and madness.

Meanwhile on the same day of 27 June 2011, *Slate.com* ran two other articles in a series celebrating homosexual "pick-up bars",[39] places notorious for promoting multiple anonymous sexual encounters and sexually transmitted diseases, complete with their own back-room bathhouses or "chapels." One of these articles included a long list of remembered "first experiences" in such bars, mostly describing seductions of teen males by older homosexual men (or women seducing teen girls). One of the confessional reports came from the flaming homosexual activist and pedophile-apologist Dan Savage, whose writings elsewhere include crude language and "in your face" homoerotic provocations, including public spitting on people he doesn't like, coating doorknobs of establishments he doesn't like with his spittle when he had severe influenza, and outbursts about wanting to "hate fuck" or murder conservative men and women who oppose "gay" marriage. He's got an entire list of other crudities on public display at his blog, in his role as syndicated "sexual advice counselor" for several porn-apologetic newspapers across the country.[40]

Surely the editors of *Slate.com* know this history, as Savage has created an uproar with his potty-mouthed and violence-promoting antics on many occasions. But for some reason, *Slate.com* was happy to publish articles extolling the virtues of older homosexuals seducing teenagers in "gay" bars, indicative of legalized pedophilia, in the very same issue where the heterosexual Reich, a strong critic of pedophilia and advocate for genitality, was slandered by psycho-drug promoter and so-called mental-health professional Kramer.

The above contradictions reveal a glaring hypocrisy in how Reich's detractors will *j'accuse!* him of all sorts of sexual impropriety and pathology which he never supported or engaged in, even as they delightfully recount or celebrate neurotic, sick or criminal sex-pathology in other articles they write or publish. This schizophrenic bias is characteristic of most of Reich's detractors, and especially for the many who published happy "reviews" of the Turner book: *leftist, uncomfortable with genitality, and approving or apologetic towards neurotic or criminal-rapine sexual behavior.*

Turner's writings have taken a similar approach, in my view, of uncritically normalizing the same sex-pathology of which he falsely accuses Reich in the "Orgasmatron" book. And it all appears to devolve down to a defense of Kinsey against Reich's social and scientific findings. Surely, if society had taken up Reich as a public scientific advisor on sexual matters, instead of Kinsey (or Dan Savage), social developments would have taken a decidedly different path, given Reich's strong emphasis upon heterosexual love and genitality, his insistence on protecting children from adult seducers, and his opposition to pornography.[41,42]

Reich Versus Kinsey on Orgastic Functions

Reich's differences with Kinsey can ultimately be traced back to their diverging definitions of orgasm, and the role of tenderness and love in sexual expression. From the very start, in his 1948 blockbuster book *SBHM*, Kinsey defined "orgasm" in the male as being equal to mere ejaculation: *"...in the present study, all cases of ejaculation have been taken as evidence of orgasm, without regard to the different levels at which the orgasms have occurred."*[43] This was not the case with Reich's clinical findings on the full-body genital heterosexual orgasm, for which the capacity for erection and ejaculation are mere prerequisites. This was detailed by Reich going back to the early European SexPol days, and repeated in the English language editions of his works, as for example in his 1942 *Function of the Orgasm*.[44] Reich always emphasized the role of tender feelings, love and intimacy for the most powerful and gratifying orgasm, which left one emotionally bonded to one's partner, and hence spontaneously impelled bonded couples towards monogamous relations, though not due to any external moralistic compulsion for a lifelong arrangement. Kinsey was aware of Reich's views on this matter, rejected them, and cited his opposition to Reich on the very same page where he proclaimed *"all cases of ejaculation have been taken as evidence of orgasm"*.[43]

Kinsey thereby confused mere *sexual climax* – which could be achieved lovelessly and in emotional isolation by masturbation, or by oral or anal intercourse – with the *genital sexual embrace* and its potentially more complete and full *orgastic discharge*. Reich distinguished between *genital* and *pregenital* sexual activity, the latter of which could build excitation and even a discharge of mechanical fluid tension, but could not provide full orgastic bioenergetic discharge – particularly for the orally or anally passive-receptive partner. The absence of full discharge drove the individual to either anxiously withdraw from sexual contact due to the unpleasure experienced, or to seek out additional new sexual encounters fairly rapidly. Undischarged sexual tension from incomplete orgasm drove people away from sexual intimacy, frequently via contactless sexual hyperactivity. This, then, was the reason for the claimed "greater sexual need" for constant multiple sex-partners as narcissistically boasted about by some hyperactive but *orgastically impotent* heterosexuals and homosexuals, an indicator of their flight from intimacy and tenderness in love.

Reich's biographer Sharaf also details these differences between Reich's clinical findings on human sexual response, and those of Magnus Hirschfeld, the Weimar-era homosexual pedophile and compiler of sexual deviations:

> *Within the reform movement, one of Reich's main opponents was Magnus Hirschfeld, a leader of the World League for Sexual Reform. Reich strongly opposed Hirschfeld's concern that the various forms of sexual [expression were] ...equally valid. What Reich advocated was a person's right to live as he or she wished sexually so long as it did not harm others. However, he did oppose a kind of "democracy of sexuality," in which all sexual expressions were "equal." Thus, he differed radically from Hirschfeld, from the decadent atmosphere of the Weimar Republic, and, indeed, from many current lifestyles.*[41]

What Sharaf references here, without being explicit, is Reich's view of bisexuality, homosexuality, "wife-swapping" promiscuity and so forth, as the consequence of neurotic orgastically impotent individuals living within a sex-confused or repressed culture.[42,45] Sadomasochism was a pathology also resultant from severe sexual disturbances, by his clinical findings, as was pedophilia, a destructive crime against children, probably committed by those who had themselves been likewise sexually abused in childhood.[41] Reich viewed the phallic narcissism of "Don Juan" characters, who boast about "multiple conquests" – as with Kinsey's uncritical reports on men with thousands of partners – as evidence of *orgastic impotence, the incapacity for full emotional surrender and "letting go" at the peak of orgastic excitation*, and not evidence of some mythological great sexual potency. Reich also viewed prostitution and pornography as the consequence of a sexually repressed, and not a sexually free society.

Overall, Reich argued in support of a more loving and emotion-based, self-regulated and non-compulsive sexuality. While erotic excitation is a centrally necessary aspect of human sexual life, Reich's clinical work showed how this does not by itself make for the most full and deeply gratifying sexual experiences. He detailed how people could "go to bed" with another and get little gratification from it; men with cold erections, women with dry vaginas, copulations with great excitation but lacking in tenderness or release, sexual embraces resembling sweaty wrestling matches (as in Hollywood films), intercourse ending with a small climax and ejaculation, "squeezed out" by tightening of muscles, but leaving the partners exhausted with empty feelings of disappointment or disgust, or with headaches or insomnia, and so on. *The capacity for bioenergetic expansion and trusting open contact with another, with emotions of tender love, must also be present for the most compete and full orgastic experience.*

What we know from Reich is that an adult can't get a full orgasm from only masturbation, nor by submission to, nor being, a compulsive sexual predator who wants only oral or anal sex, nor by promiscuous multi-partner empty-sex "fucking". Sitting in an orgone accumulator won't do it either.

I mention these facts because they so clearly place Reich in opposition to Kinsey and his supporters on fundamental issues, and therefore appear central to understanding Turner's "Orgasmatron" attack on Reich. Reich remains, as always, "uncool" (emotionally warm), pushed out of the public discussion except when being mercilessly distorted and attacked, by the very same people who endorse Kinsey's "cool" agendas.

Before any of what I relate above was publicly known about Kinsey, *Time* magazine actually put his picture on the cover of their 24 April 1953 issue, effectively promoting his "anything goes" empty-sex discharge theory, where love was basically a side issue, or irrelevant. But they, like most other mainstream media, never had anything friendly or accurate to say about Reich. *Time* editors would repeatedly slander Reich, eventually to dance on his grave, even into the 21st Century, as detailed below.

I will conclude this section by restating how Reich's influence during this period of American social transition was limited, and then only towards authentic sexual and social freedoms, underlain by a self-regulated and non-compulsive morality founded upon heterosexual genitality and love. He emphasized love, intimate emotional bonding of couples, and protection of children from abuse and sexual predation as center-most principles. He was surely critical of the rigid church doctrines of his day, which forbade all sexual pleasure outside of marriage, as well as prohibiting contraception, abortion and divorce. But Reich never supported the sexual license and porn which spread across the landscape primarily under the banner of unscientific Kinsey poly-sex ideology. These important distinctions are hideously distorted by Turner, exactly on the points where Reich and Kinsey diverge.

Reich and the "Bohemians"

CLAIMS (p.6): Turner states: *"Orgone boxes were used by such countercultural figures as Norman Mailer, J.D. Salinger, Paul Goodman, Allen Ginsberg, Jack Kerouac and William Burroughs – who claimed to have had a spontaneous orgasm in his. ... Bohemians celebrated the orgone box as a liberation machine, the wardrobe that would lead to utopia, while to conservatives it was Pandoras box, out of which escaped the Freudian plague – the corrupting influence of anarchism and promiscuous sex."*

FACTS: Here Turner once more rattles off the names of people he claims are representative of Reich's work and ideas – none of them are, however. Each of them promoted their own different brands of sexual freedom, licentiousness, or neurotic perversions. If they mentioned Reich or happened to use an orgone accumulator, it was without accurate reference to his ideas.

In this section, Turner also repeats some of what he declared in his 2004 article, as mentioned at the start of this Chapter, which in turn was apparently obtained and updated from the dead smear-writers Brady[9] and Gardner.[10] He defines Reich by inaccurate casual references to alcohol-saturated or drug-addled *freedom peddlers* (as Reich called them), including "hipster" bisexual "swingers" and pedophiles, as if the fact that they occasionally sat in an orgone accumulator, or wrote something kindly about Reich actually proved something. Well, it does prove two things:

1) The accumulator may give people a bit of a sensible boost in energy, even if they are Bohemians, but ...

2) *The orgone accumulator* <u>cannot</u> *cure alcoholism or drug addiction, nor make people orgastically potent, nor provide mental health.* And that, Reich *did* assert quite clearly (some quotes given below).

CLAIM and CLARIFICATION (p.7): Turner gives a long and bizarre quotation by James Baldwin, another drug-abusing anarchist homosexual writer who, with some remorse, reflected back upon his times of *"experimentation with sex, with marijuana, with minor infringements of the law"*. Turner provides this quotation as a means for blaming the dead Reich for Baldwin's sorry situation, merely because someone in Baldwin's circle had an orgone accumulator. Again, this proves the orgone accumulator cannot make people sexually or mentally healthy. But a more critical point which Turner misses is, that people like Baldwin came to their point of (temporary?) remorse and self-reflection only after having firstly exhausted themselves with all the possibilities of empty-sex and brain-fogging drugs. Then they hit bottom, and realized it. But it would be a stretch to claim that Baldwin had any clear appreciation for Reich.

The last sentence in the Baldwin quote provided by Turner says it all: *"The people I had been raised among had orgasms all the time, and still chopped each other up with razors on Saturday nights."* No, these were not "orgasms" as Reich defined it, but merely sexual climaxes, empty-sex discharges in the Kinsey style.

The pent-up emotional tension and angry rage which those people carried in their guts, inside their character structure, and which led them to "chop each other up", also worked to block their capacity for full loving surrender and intimacy in the sexual act. *"Discharge" alone has little or no ameliorating effect upon biophysical condition or character structure.* If it did, then the Imperial Japanese soldiers who raped so many "comfort women" during WW-II would have been "comforted" and spontaneously stopped their killing machines. Likewise with the polygamous Islamic-Sharia culture breeding terrorism, where plenty of "screwing" goes on, with slave women or little girls from arranged loveless marriages, or by sodomizing boys – but they never gain sufficient release from bottled-up inner tensions, nor subsequent mental clarity to stop hating and murdering the nonbelievers. For untold generations, the most violent of Saharasian cultures have had plenty of "discharges" via loveless rapine screwing, with no softening benefits to character structure, and won't until the day comes when they allow their children to grow up unmolested, and treat their own women as equals, with a humane and loving attitude.

That's the message from Reich, bound to infuriate a lot of people who don't want to hear it, and utterly lost on people like Brady, Gardner, Baldwin, Burroughs, Kerouac, Ginsberg and Turner.

By page 8 of Turner's Introduction, he takes several additional cruel and off-the-mark shots at the Summerhill School in rural UK, as a follow-up to his 2004 missive.[2] Why should he hate this school so much that it obviously preoccupies him? Is it because at Summerhill he observed a group of truly self-regulated and happy heterosexual kids?

Turner's text drags on to selectively quote from Reich's ex-wife Ilse Ollendorff, who gave an unclear description of Reich's emotion-release therapy methods, by Turner's recounting in any case. Turner's discussion at this point in the book adds nothing to clarify things, and thereafter he goes off into a muddle.

More Nasty Accusations: Madness and Porn

Turner provides very few quotes from Reich, but he is sure to emphasize the ugly words from some of the old European psychoanalysts who hated Reich's guts and spread false rumors of his "going insane", as on prior pages where Sandor Rado is cited. The implication is, *because Rado says it, then it must be true!*

In fact, Rado and many of the psychoanalysts felt deeply uncomfortable with Reich's new definitions for *sexual potency,* and with his arguments that childhood sexual latency and the Oedipus conflict were culturally and not biologically determined.[53] They either could not fathom, or loathed his clinical findings on orgastic disturbances and genitality, as well as his divergence from psychoanalytic free association and other changes in therapeutic technique. And this was well before Reich began to examine people's bodies clothed only in undergarments, identifying such things as skin pallor or shallow respiration, and actually began pushing on their stiff jaws or rigid chests to release the pent-up emotional energy trapped within their somatic armoring.[18]

Reich's efforts to merge Freud and Marx also discomforted some, but not all Freudians, as many were already far more "pure Marxist" than Reich ever was. Reich's public criticisms of the Nazis were viewed as highly provocative also, given the general appeasement mentality within mainstream psychoanalysis (*"Don't provoke them!"*). His early 1930s books *Mass Psychology of Fascism*[46] and *Sexual Struggle of Youth*,[47] advocating the naturalness of heterosexual love between adolescent peers (ie., Romeo and Juliet), was the final blow for nearly all within German psychoanalysis. "Reich is insane", the rumor mongers whispered. The Marxist groups also thereafter complained bitterly about Reich's ideas, especially when he dared to criticize the Communist Saints and urged sexual reforms over class struggle. They publicly denounced Reich and expelled him from the German Communist Party, just as Hitler came to power. Reich was soon on both Gestapo and NKVD death-lists, as detailed previously.

Such clarifications aren't necessary for Turner, however, and he adds mud to the water at this point in his narrative by zipping forward from the 1931 opinion of Rado some 20 years to Reich's 1950s observations and photographs of UFO flight-tracks, contained in one of his last books, *Contact With Space*.[48] Turner also ridicules Reich's cloudbuster invention, as if its mere unusual nature automatically made it unbelievable and Reich a crazy person. (The dear reader may wish to pause a moment to consider how the telescope, microscope, steamboat, automobile, airplane, and rocket were initially greeted with ridicule and outrage by the people of their day.) Turner's inference here is that, because Reich saw UFOs and engaged in weather experiments, then "he must be crazy". That presumes everyone who has seen a UFO must "be crazy" also, including several former US Presidents, a large number of military and airline pilots, astronomers, astronauts, and people from all walks of life. And all the scientists engaged in weather modification experiments also "must be crazy"?

By page 13, Turner quotes from a 1964 article in *Time*,[49] which blamed Reich for a claimed widespread pornography and promiscuity they had identified. Apparently the editors of *Time* and Turner both forgot about Kinsey being on the *Time* cover only 11 years earlier, as well as Kinsey's massive media blitz and political influence. But that would not be the first instance where *Time* would attack Reich, nor the last. They slandered him in articles[50] dating back to 1947, which was shortly before they celebrated Kinsey, and "coincidentally" in the same year the Brady smear campaign began. *Time* has been as regular as clockwork in making periodic smears and jabs at Reich, all parroting the 1947 Brady lies. In their pre-millennium edition of early 1999, *Time* magazine fully skewered Reich, along with Pons and Fleishman of the cold-fusion experiments. In an article on *"Cranks, Villains, and Unsung Heroes"*,[51] they put Reich's photo on the same page with Stalin's Lysenko and Hitler's Mengele.

The depth of hatred and willingness-to-lie as was necessary to compare Reich to Lysenko or Mengele revealed the foul character of the left-wing editors of *Time* magazine more than it said anything about Reich,[52] a man who had risked all to openly fight the Nazis and later the Stalinists, and who had been dead since 1957. We may ask wherefrom springs this insane obsessional emotional rage, for writers and editors to behave like the old Bishops of Rome, who periodically dug up the rotting corpse of some long-dead heretic, lashed it to a pole and burned them posthumously in a public spectacle. One hopes the American and European public will eventually see through such spectacles, though "eventually" may take a long time indeed. Many Letters to the Editor were sent to *Time* by professionals in objection to that last turn-of-the-century slander. I know because I wrote one myself, as did many of my professional associates – none were published.

Turner gives no hint of this on-going malicious hatred of Reich by so many in the mainstream "intelligentsia" and literary crowd. Neither does Turner give Reich any credit for having contributed to the softer and more rational-healthy aspects of the sexual revolution for which he truly advocated – as with his calling for legalization of contraception and divorce, an end to severe prudery about the naked human body, the rights of young unmarried couples to have their romances and a premarital love life without going to prison or committing suicide, his push for better care of infants and children, with protections for pregnant and nursing mothers, and the improvement of the rights and status of women all around. Today we have all this, more or less, at least in Western society, but at the time of Reich's European SexPol work, this was not the case.

Reich's SexPol platform from the 1930s is eminently sensible and contains much that conservatives would agree with – and which opinion polls suggest, take advantage of, as do most other people in the liberal Western democracies. Divorce and contraception are accepted by most Americans, including by most American Christian conservatives, even as the Popes and many preachers might speak against them. Abortion is also legalized in the Western democracies, and arranged marriages exist only among certain immigrant communities, such as Hindus, Chinese or Muslims. And in the majority of cases where young women are killed for "crimes" of breaking virginity taboos or "disobedience", it is overwhelmingly at the hands of Muslim fanatics, whom the liberal-left pro-Kinsey intellectuals are always rushing to defend, and gushing over.

But these same "literary intelligentsia" give Reich all the blame *for things he in fact opposed* and which their own radical leaders approve of, such as: the explosion of public porn and fantastically promiscuous fucking in both hetero- and homosexual sex-clubs, or narcissistic phallic obscenity in public homoerotic/pedophilic "street festivals" and "parades", characterized by pubic nudity and open screwing, S&M flagellant exhibitionism, with drunken screaming, defecation and urination (i.e., the "Folsom Street Fair", "Berlin Love Parades", etc.).[53] These madhouse affairs continue only because liberal "intellectuals" have managed to successfully defend them in the courts, with a push to "normalize" such monstrosities as pedophilia, bestiality and sadomasochism as "alternative sexualities". Hollywood and media types join in with the social degeneration by celebrating little girls dressed up as adult sexual objects, and by conning confused boys and young men into "metrosexual" lifestyles, where they become fodder for pedophiles. Young girls also are transformed by starvation fashion-chic into looking like skinny boys, typically by homosexual fashion designers who, we are told, "love women". A major theme in all such collective neuroses is the sometimes concealed, sometimes quite open hatred and fear of heterosexuality – *a hatred and fear of the opposite sex, and not "love" of the same sex* – with a push to declare the sexual confusions of youth (created by confused-repressive or licentious adults) as "natural", something "needing adult supervision" or horrific sex-change surgeries to "correct". Such is the background social characteristic of those who attack Reich on the one hand, while on the other, publicly promote pornography and "pedophile science".[54] (Also see the Appendix.)

My study of these socially pathological factors suggests that much of liberal Western democratic society is being gradually led and shoved away from its essential and basic foundations upon heterosexual love, away from Reich's affirmation of healthy genitality, and towards a Kinsey/1960s poly-sex pedophile agenda, with some kind of weird feminist embrace of polygamy and the Islamic veil to boot. Forced marriages of young girls to old men, man-boy pairings, female genital mutilations, and a push to erase freedom of speech and press under new "hate crimes" codes, also are entering into the West under this co-related umbrella of the modern illiberal leftist embrace of politically-correct pornography, polygamy, homo-pedoism and inconsolably violent Islamic extremism.

While such things are neurotic and psychopathological by the Reichian view, Reich often gets publicly blamed for them by those who dominate the media microphones and publishing industry. Or alternatively, Reichians and moderate Americans in general are labeled "prudish phobes" for objecting to the "hip-swinger" trends. Turner has little or nothing to say about these kinds of social horrors which come to us thanks to Kinsey and the emotional-plague cult of pedo-oriented poly-sex "social activism", which he seems to support.

Thankfully, this behavior and self-blindedness does not afflict every liberal or homosexual, but does exist within their organizations and demands to be recognized as such, dealt with firmly, and stopped. Many know the truth about all that I write above and are appalled by the situation. However, out of ignorance of what to do about it, or from fear or habit, few within these social networks dare to speak out against the trends, lest they be falsely accused of being a "right-winger", "racist", "homophobe" or "Islamophobe", and then socially isolated, or even verbally or physically attacked.

Turner's book utterly fails in this regard, to educate and warn the public about who is really responsible for the current social-sexual degeneracy, where it does actually exist. Nor does Turner educate about the more healthy aspects of the sexual revolution. He reveals no appreciation about the severely repressive conditions which existed across most of Europe and Russia before and during the Nazi-Stalinist times, when Reich was active in sex-political work. Nor does he detail in his large book the comparatively free social-sexual conditions throughout the Americas, including or especially within the USA, over most of the 20th Century.

American Conditions and Sexual Freedom

Turner's "Orgasmatron" book carries the strong odor of a well-worn "Euro-Superior" theme, which demands at least a short response. Firstly we should consider that no nation on our small planet today fulfills the requirements of a truly sexually healthy society. By the early 1900s, only a few isolated tribes managed to avoid the sweeping social conflagration of invasions and wars over the last 6000 years, which destroyed what remained of our more peaceful human cultural patterns. This process was generally understood by Reich, who pointed to the Trobriand Islanders of the early 1900s as a decent representation of a sexually healthy and non-violent society,[55] fulfilling his sex-economic predictions. My *Saharasia* work provided more detail on that question,[56] comparing and contrasting over 1100 different world cultures against each other. While both America and Europe have a nearly equal and large sexual freedom today, protected by law, both also have an inordinately high percentage of sexually impotent neurotics, along with much sexual misery and resultant social chaos. This is seen in the high levels of unhappy marriages, drug abuse, alcoholism, child abuse, suicides, consumption of psychiatric drugs, and left/right political extremism and intolerance. But I do believe a comparison of trends places America on the healthier side of the equation as compared to Europe, which has over millennia been badly affected by extremist cultures found within the adjacent regions of Saharasia – a factor which by geographic distance alone has allowed Americans to develop along more self-regulated and freedom-oriented pathways.

As Europe relaxed in the tenuous peace between two major World Wars, America had the roaring 20s and the early Hollywood cinema which in a healthy manner celebrated youth and love, including some natural nudity, but without the porn-pathology as is so widespread today. There were rational social movements for women's rights, aiming to provide women with the technical means and legal power to control their own fertility and destiny. The American sexual reform movement was in some measure more successful as compared to similar European movements, which were nearly always opposed by the various Emperors, Kaisers, Tsars, Kings, Führers and Communist dictators. Such Royals or dictators simply did not exist upon the American social landscape, and Americans had the advantage of being able to elect new political

In Defense of Wilhelm Reich

leadership every two years, at national, state and local levels. By contrast, and with the notable exception of the Western European democracies, Kings, Kaisers, Tsars and Führers had to be petitioned to "pretty please" do this or that to help the ordinary people, which they frequently on whim either did or did not do. Even today, Europeans are generally treated like children by their EU "parents" and the various political party bosses, who feel they "know what's best", and won't even allow voting referendums on such life-changing concerns as with the massive EU Constitution and its nanny-state micro-regulations of love, work and knowledge.

American women also generally had greater freedom than most of their European cousins for most of the early 1900s. World Wars I and II led additional young women out of the kitchen and into factory or farm jobs, and eventually into the universities, to take up the slack with the young men gone to fight. Many young American soldiers – who already had greater sexual, economic and political freedom than most of their European cousins – when on station were openly romanced by young women of Europe, Australia and the South Sea Islands, and the young men were happy to accommodate, often marrying such women. No coercion was involved as in the Axis-controlled areas. American soldiers kept the "pin-up girl" images nearby, an expression of their healthy heterosexual longing. They painted beautiful girls on their airplanes, similar to how the defensive-only military of the peaceful Trobriand Islanders decorated their war-shields with breasts and vulva, something which probably unhinged the woman-hating Dobuans who periodically raided their islands for booty and slaves.[55,56] German and Soviet aircraft had no such lovely images, only the Swastika, Iron Cross, Red Star, and Hammer and Sickle.

Americans and the Western democratic allies did not value the naked muscular homoerotic male imagery, or the worker-bee desexualized female imagery as seen in the statues of the Third Reich or Soviet Union. And those same Western democracies never forced women from liberated Axis territories into prison brothels, for rapine abuse by their soldiers – as was widespread for Nazi and Imperial Japanese soldiers. No death-camps were ever created by Americans (or allied British) either. Nor did American (or British) troops ever commit such barbarisms against civilian populations as when the German Wehrmacht swept across Europe, or when the Red Army entered Eastern Germany, and Soviet generals criminally allowed their soldiers to exact vengeance against German women and girls in massive rapes.

In the lead-up to the Normandy invasion to liberate Europe during WW2, when American GIs flooded the British Isles, the slightly more reserved British complained, only half-joking, that the *"Yanks are overpaid, oversexed and over here!"* Complain as much as the Europeans did in the postwar period, the social customs, music and dance from the plain-talking "big-toothed American" farm-boys and city-dudes flooded Europe after those World Wars, and their mannerisms, music, radio, films and dance were eagerly absorbed by European youth all over. Americans were the descendants of those who had thrown off the shackles of Kings generations before the Europeans. American women also pushed the limits, for the vote, for contraceptives, for equality in marriage, and other things. And this blended seamlessly with the earlier Anti-Slavery movement, and the later Civil Rights movement. Little of this owed any direct debt to Reich, but his work was in harmony with it, gave it support, and was a part of it. His publications sometimes described the healthier contrasts between the "mongrel" American versus the rigid pureblood goose-stepping "Übermensch Aryan", between the free-wheeling big-band-dancing girl-chasing American versus the stiff and moralistic "Soviet proletariat".[57]

While America surely had and has its problems, and its own share of rigid prudes, it was never as repressive as most anywhere in the Old World one might point to for a given period, especially as compared to the Islamic regions where independent-minded or sexually-active women, as well as young homosexual males (but never old pedophiles) continue to be legally murdered by public stoning, beheading, or being hung from cranes. Nor was America ever as repressive as Europe under the Kaisers, Tsars and Kings of the 1800s or 1900s, nor ever as chaotic as in Weimar Germany or post-Tsarist Russia under the Provisional Government, where various pro- versus anti- political parties rushed around madly shooting each other dead in large numbers. America was light-years away from the conditions seen under Hitler, Hirohito, Mussolini, Franco, Lenin, Stalin, Mao Zedong, Pol-Pot, Mugabe, Ho Chi Minh, Castro, Khomeini, King Saud, or any other right- or left-wing, or Islamic dictator one might name from the 20th Century.

And yet, Americans are continually ridiculed or slandered by self-ordained soapbox preaching Marxist-revisionist "historians" of

In Defense of Wilhelm Reich

the hard-left, in both Europe and America, claiming American freedoms are a sham, that we here in "the colonies" are all oh-so sex-repressed as compared to "liberated" European socialists, and so on. But in fact, *this is a falsehood*, and left-wing Europeans looking down the nose at Americans is typically an old and compensatory emotional habit, going back hundreds of years.[58] *It was the Europeans, after all, who invented Marxist Communism, Hitlerite National Socialism, and Mussolini Fascism as "answers" to their respective emotional and socio-economic problems* – and we all know how well that worked out! The Euro-left also habitually points to the French Revolution as a model, but never to the American Revolution which preceded it, and had already broken free of most feudal tendencies. Nor was bloody terrorism invoked by American Revolutionaries against British loyalists in retribution, as one saw with the busy guillotine in France.

I've written entire chapters in my *Saharasia* book[56] devoted to similar issues, of regional variations in sexual freedoms and family life. There were many notable ancient societies and aboriginal tribes around the world with a high status for women, great social and sexual freedom, potent herbs used for contraceptive purposes, and similar things. And as a consequence, they also had very cooperative and peaceful societies, with low levels of social violence. I've also detailed how every kind of sex-hating religion made war against contraception and abortive herbs, and tried to kill off the women who had such knowledge. The burning of women at the stake by the Catholic Inquisition was frequently for the "crime" of having contraceptive or abortive knowledge. Much of the struggle up from the nightmare of the Inquisition and other theocratic dictatorships – through the Reformation, Renaissance, Enlightenment and Democratic revolutions against the Divine Power of Kings – is a history of similar struggle for sexual, reproductive and marital freedom, of people seeking to control their own life and love destiny. *Reich was a part of that struggle, and his SexPol activities and writings did not constitute any endorsement for the retrograde social changes stimulated by Kinsey and friends, and which Mr. Turner accuses him of and blames him for.*

The Turner book continually rehashes the points I address above, so it would serve no purpose to recount those repetitions. But there are a few real zingers to be found in the later parts of his book which are so outrageous as to demand additional specific rebuttal.

Reich's Bioelectric Experiments, and the Lie About "Masturbation of Patients"

CLAIMS: (p.178) *"Whereas in Reich's laboratory experiments subjects were encouraged to masturbate, his therapy was entirely different. He did aim to improve orgastic potency but this was done through massage of other parts of the body, in the hope that such treatment would free muscular blocks. The distinction was important to Reichians – though of course to their critics the line seemed extremely fine – because actual 'masturbation therapy' was at the time a competing school of thought."*

FACTS: This outrageous proclamation is loaded with dirty-minded innuendos. Firstly, it falsely suggests Reich's laboratory experiments constituted some kind of "masturbation festival". In fact, his bioelectric experiments were carried out in a laboratory room at the University of Oslo, Department of Psychology, and addressed the issues of both emotional and sexual excitation. His book on *The Bioelectrical Investigation of Sexuality and Anxiety*[6] gives the full details, and in it one will find a recounting of the very first scientific measures, via bioelectrical recordings, of human sexual and emotional excitation. One report is given of a test subject who did masturbate, but the other experiments focused upon more mundane things, such as a couple kissing, or the pleasure experienced from a warm gentle stroking of the hand by a trusted companion, compared to a hard stroking. In another test, the bioelectrical tracings recorded reactions of a person eating a spoon of sugar, or recorded the "disappointment reaction" when told they were being given sugar but instead were given salt. Tickling versus pressure reactions were recorded. Or when a kiss became unpleasant. In another, a relaxed subject was startled by the popping of a balloon. These experiments produced the first quantitative measures of emotional phenomena, previously only subjectively described. They also addressed larger and centrally important questions about pleasurable versus anxiety-provoking or unpleasurable stimuli.

The overall discovery was, firstly, that erogenous tissues have a generally higher charge and a more reactive bioelectrical characteristic than other skin surfaces. Secondly, Reich documented how pleasurable experiences produced a bioenergetic expansion with an increased electrical charge at the skin surface, while anxiety created an overall bioenergetic contraction with reduced skin surface charge. This was an important breakthrough finding, especially as regarding how, under conditions of chronic stress and pain, the sympathetic branch of the autonomic nervous system could become reactively habituated, resulting in chronic anxiety reactions and a blocking of parasympathetic pleasure functions.

Reich's experiments thereby documented the bioenergetic aspects of pleasure versus anxiety, as well as the differences between healthy people's capacity to expand out into the world, towards meeting of needs and desires of all kinds, versus the big problem of life-trauma leading people into conditions of pleasure-avoidance, frigidity, sexual anesthesia, panic, and chronic anxiety. His work clarified many questions on human sexual response, such as the widespread incapacity of armored humanity to surrender in the sexual embrace, which invariably blocks sexual excitation and the path to full orgasm. Reich's work on these subjects touches upon emotional and sexual issues which the average person struggles to suppress and avoid, and that includes those health professionals and biologists who affirm the "normality" of sexual dysfunction and emotional retardation, and often do little more than prescribe drugs to help people avoid their core problems. Neither Kinsey nor Masters and Johnson, nor any of the major "sex researchers" since, have touched upon the basic issues which Reich elucidated in both sociology and biophysics. His Oslo experiments were therefore not some kind of "jerk-off" extravaganza as Turner and other Reich-critics suggest, but rather constituted the first electrophysiological measures of basic emotions, sexuality and neurology.

We can determine the facts by referencing Reich's own publications, wherein he identified a rumor campaign launched against him by the "psychoanalysts and neurologists" of his day, as in this quote from a longer published report in one of Reich's research journals, by Dr. Alan Cott:

"...in the Spring of 1948, a group of psychoanalysts and neurologists decided to smash Reich's work. ...several obviously malicious attacks, unworthy of physicians, have been launched [to include] Sudden outbursts of malicious gossip in mental institutions and in university lectures to the effect that patients are "being masturbated" in orgone therapy. We have many legally executed testimonies of witnesses regarding this

concerted campaign of defamation of orgonomy. We have transmitted all documents to our law office for eventual legal prosecution of these criminal gossipers who, instead of helping to solve the misery of children and adolescents, are busy with defamation, slander, lies, distortions. ...

    A. ALLAN COTT, M.D., Secretary
    Committee on Medical Orgone Therapy" [59]

Here's another statement by Reich's ex-wife Ilse Ollendorff, written some years after Reich's death:

"One thing has to be stressed, that vegetotherapy or Orgone Therapy has nothing to do with mechanical massage or with 'masturbation therapy.'" [60]

While Reich did write that masturbation is a natural development for children, and as a temporary measure is not harmful for adults (except if it becomes a chronic means for avoidance of the genital embrace), the accusation that Reich "masturbated his patients" is entirely false. From Reich's publications and archives, however, the source of this salacious rumor was identified, as alluded to above, in the quote from Dr. Cott. In fact it came from within the circle of Freudo-Marxist and Nazi-apologist psychoanalysts who had conspired to throw Reich out of the IPA in 1933, when Jews were being expelled.

Those and other malicious fabrications appeared in Fenichel's *Rundbrief* letters, which were mailed in multiple copies to all the top psychoanalysts of that period of exile during the war years, keeping everyone abreast of new goings-on. In them, Fenichel frequently and openly disparaged Reich, directly calling him a "madman", or repeating similar statements by others, alluding to claimed "personal foibles", attacking his new scientific works, his therapy methods, calling him a quack, and so forth. The details on this are now published in two massive volumes *Otto Fenichel: 119 Rundbriefe*,[61] which is one of the most important uncensored historical documents revealing private discussions within the inner sanctum of international psychoanalysis during the period from 1934 to 1945.

The malicious gossip Fenichel circulated in his *Rundbriefs* was further spread by their recipients in whispering campaigns in

In Defense of Wilhelm Reich

Europe and notably within America, especially when Fenichel worked for a time at the Menninger Clinic in Topeka Kansas. There, Fenichel filled the ears of Karl Menninger and other staff members with poison about Reich. This would later figure into the American smear campaign in the USA, when Menninger reprinted one of Mildred Brady's smear articles attacking Reich and the orgone accumulator with salacious innuendo, in a 1948 issue of his *Bulletin of the Menninger Clinic*.[9]

Two German-emigre psychoanalysts, Hermann Nunberg and Annie Reich-Rubinstein (Reich's ex-wife Annie Reich, remarried to the Comintern Marxist Arnold Rubinstein) were also identified as spreading rumors about Reich allegedly "masturbating a female patient." When this outrageous lie came to the attention of Wilhelm Reich, his ex-wife and Nunberg were confronted by Reich's lawyer. The woman patient was eventually identified and questioned, but she and the two psychoanalysts denied ever making the accusation. As it turned out, this woman had never met Reich, nor had she ever been in treatment with any of his trainees or associates.[62]

Members of the Consumers' Union (CU – then under Communist control) also got involved in spreading slander, as when Dexter Masters (who became CU Executive Director), spread another set of rumors, declaring he was *"...a friend of Miss [Mildred] Brady"* and that the orgone accumulator was *"a box whose purpose was to activate masturbation."*[62] In fact, Masters was a one-time lover and close associate of Mildred Brady, who then ran the Western Division of CU. Reich's participation at the Lucerne 1934 IPA Congress was cited as background evidence for this accusation, where Brady and Menninger had both been in attendance. That conference was held years before Reich's discovery of the orgone energy or his invention of the orgone accumulator, but no matter. Brady's malicious smear article not only appeared in the 1947 *New Republic*,[9] but later within the 1948 Menninger Clinic *Bulletin*,[9] with shorter condensations the next year in both the *Journal of the American Medical Association*[63] and *Consumers' Reports* .[64]

None of these publications made any kind of fact-check or investigation as to the truth or falsity of the Brady accusations, or of the rumors, as they were all friends or associates, with shared hatreds of Reich based upon psychoanalytic or Marxist ideology, or both. Together they worked to fabricate new "truths" about Reich's claimed "insanity" and "masturbating patients", which were spread even further by professional gossip and innuendo, and in various copycat articles.[65] All of the above primary articles were forwarded to the FDA by physicians hostile to Reich, with encouragements to "investigate", which they eventually did with great prejudice. These same slanders were later repeated within the US Supreme Court's decision against Reich, as part of the "evidence" they reviewed when making their legal determinations to not hear his case on appeals! And today, of course, *we have Mr. Turner and friends to thank for spreading this gossip-slander even farther!*

The reasons for this severe hatred of Reich by the predominantly Freudo-Marxist psychoanalysts appears at root to be a personal nervousness about the new steps Reich had been taking, as I give above in the section about Sandor Rado proclaiming Reich insane. But there was another element as well, in that Reich remained highly respected in some quarters of psychoanalysis for his innovations in therapy, which actually helped people far better than the older mode of free association on a couch, where you don't even see the analyst (who can then conveniently take a nap).

After fleeing Germany, Reich was warmly received by the Scandinavian psychoanalysts, some of whom went into training with him. Other psychoanalysts remaining loyal to the IPA considered this a threat to their organizational existence, if not also to their personal incomes. The same can be said with respect to the threat posed to the medical establishment and pharmaceutical industry by Reich's drug-free therapy methods, and the immune-boosting orgone accumulator. These crass economic considerations constituted an additional motivation for his enemies to try to destroy Reich's professional standing.

Fenichel and other Marxists within the psychoanalytic movement had even more personal reasons to fabricate and spread nasty rumors, based upon Reich's growing criticism of the Soviet Union and "pure Marxism". Fenichel remained a "Red" to the bitter end, as did many other Freudo-Marxist psychoanalysts, including those such as Erich Fromm and the Frankfurt Group, which had fled to London. They considered Reich to be "a bad Marxist" for example because of his attribution of human irrationalism to emotional trauma, genital frustration and character structure, rather than to Marxist historical materialism and class-conflicts. Worse, to the Communist Party leaders, Reich wrote that neuroses and sexual impotence were found just as frequently among the Marxist proletariat as among the counter-revolutionary bourgeois, which contradicted Marxist-Leninist thought entirely.

It is therefore ironic that today, Reich is denounced by a few ignorant conservatives[66] as a central figure of the Freudo-Marxist *Frankfurt Group*, to which he *never belonged,* and whose membership accepted Reich's ideas only nominally, viewing him as *insufficiently Marxist.* The Frankfurters and the *New Left* which arose in later years primarily cite Reich's very early Marxist writings, which he later heavily edited or retreated from. His later more conservative writings on sex-economic and orgone biophysical subjects are ignored or denigrated, by contrast, as they frequently include his anticommunist discussions. The Freudo-Marxists and New Left also frequently believe Reich "went insane" in the late 1930s, after he was expelled by the *German Communist Party*, and took up orgone energy research. They also tend to follow the Kinsey line on sexual behavior.

Reich's ex-wife Annie Reich-Rubinstein, who remained a loyal mainstream psychoanalyst, if not also a Marxist, absorbed these slanders completely, and in the years thereafter had nothing decent to say about her ex-husband. She not only spread her own malicious lies against Reich, but tried to poison the minds of their two children, Lore and Eva, both of whom were educated as physicians. Eva Reich later in life saw through her mother's hatred and became a close participant in her father's work, teaching professionals and laypeople about the orgone accumulator and cloudbuster, and training midwives in her *Orgonomic First Aid* and *Butterfly Touch Massage* methods[67] for helping distressed infants. She became a public advocate for gentle home-birth and argued to end genital mutilations (circumcision). By contrast, Lore Reich followed in her mother's footsteps to become a conventional psychoanalyst, never getting involved in her father's work. In her elder years she began parroting the same deceit and poison as she received over decades from her angry mother Annie Reich-Rubinstein, and other psychoanalysts. Turner predictably quotes Lore Reich on such matters, but without a hint of the historical issues as I give above.

Silvert's Betrayal of Reich

CLAIMS and CLARIFICATIONS (p.316-324): There is a known case where one of Reich's medical trainees, the obstetrician Michael

Silvert, without Reich's knowledge, was seducing a female child therapy patient. Turner covers this incident, but convolutes it into a weapon against Reich's entire life-work, misrepresenting it with the false suggestion that Reich might have approved of pedophilia. The girl, "P.W.", is now a grown woman and so full names won't be given, but in the Turner book she is quoted:

> *"I found out many, many years later that Silvert did tell Reich what he was doing, or I guess it came out in his therapy with Reich, and Reich flipped out. It did redeem Reich in my mind a lot; I never thought he was responsible... never ascribed Silverts behavior to Reich, it was an aberration."* [68]

Turner reproduces this quotation, which accurately indicates Reich did not know what Silvert was doing, and reacted with appropriate anger and actions against Silvert once he knew what was going on. Turner noted that Reich forbade Silvert from treating women patients and suspended him from practicing orgone therapy for a year. That latter information came from the memoirs of Elsworth Baker, a physician who also worked with Reich.[69] Personally speaking, I feel Reich was too optimistic about therapeutic treatment for adult sexual offenders. While Reich saved the young woman from further predations by Silvert, his decision to continue working with him on other matters eventually would prove deadly for Reich. In some kind of psychological revenge, Silvert later was responsible for knowingly violating a federal court injunction which had been imposed upon Reich by the FDA, against moving orgone accumulators and books across a state line. Even though Reich was at the time working in Arizona and knew nothing about it, Silvert's action in shipping a truckload of books and some orgone accumulators from Maine to New York triggered a legal reaction against both men for violation of the prior FDA injunction. Ultimately both men landed in prison – leading to Reich's death, and some months later, to Silvert's suicide.

Turner's admission of Reich's correct actions to protect the girl, by demanding Silvert no longer conduct therapy with females, was a gracious but temporary lapse from habit. Move from page 321 to 322 in Turner's book, and he once again besmirches Reich's efforts to allow unmarried youth comparable to "Romeo and Juliet" the rights to their healthy romance (and thereby to *not*, like those desperate young lovers, commit suicide). Turner's statements here are an amazing bit of hypocrisy, when compared to the uncritical, even fascinated presentations he gives elsewhere regarding homoeroticism, Molly Houses, and Kinsey.

Turner then presents a report about one of Reich's trainees, Albert Duvall, who – *years after Reich died* – seems certain to have engaged in reprehensible scare tactics with child therapy patients, in efforts to get them to scream and cry. But the Duvall incident is reported by Turner with suggestions that it involved child sexual abuse, which is not established. Turner quotes from an interview with Morton Herskowitz, another Reich-trained physician, whose statements about Duvall and sexual abuse – as reported by Turner – sound rather off-the-cuff and difficult to accept at face value. I also have a separate report from physician Richard Blasband, who recalled the Duvall incident from his days working within the same group of Reich's trainees, and is certain there were no sexual improprieties involved.[70] Additionally, two of Duvall's primary accusers and former child therapy patients, "L.L." and "R.W.", have published recollections detailing Duvall as a heavy-handed and incompetent child psychiatrist, but nothing more. They remained justifiably very upset about the scare tactics they were subjected to as little children. However, they did not report any sexual abuse.[71] "R.W." also has been attacking Reich's orgone biophysical work on internet for many years, in the manner of the "skeptic" clubbers, and just like them, he chronically lies by omitting mention of the many verification studies on Reich's biophysics. Another of Duvall's critics, "M.B.", likewise blames Reich for his sufferings under Duvall's bad therapy and claimed sexual abuse, years after Reich was dead. But "M.B." is no innocent to be trusted. Aside from chronically slandering Reich, "M.B." *publicly endorses and admits to bestiality – human-dolphin sexual relations specifically, but not limited to only the wet species – in a published book.*[71]

What a mess!

These strange men routinely and publicly lie about Reich's orgone biophysical work, take up the slander-ridicule language of the "skeptic clubs", and use their bad experiences with Duvall as a means to attack Reich, even though Reich was dead for many years when the abuse (or claimed abuse) happened. Simply put, *they are not reliable informants.*

There was one additional case of a woman "S.S.", who at age 60 remembered being molested when 3 years old by a female therapist who was temporarily moving in Reich's orbit. The problem with this report, beyond the issue of great memory distance, is how she joined with "M.B." and "R.W." to excoriate Reich personally with "skeptic club" language, in a chorus to blame him for their unhappy lives. We are therefore confronted with questionable claims from all except "P.W." whose recollections about Silvert appear more balanced and reliable, probably because she was helped by Reich's strong intervention against her abuser.[71]

Turner, who possibly knows of this problem, admits on the same page 322, *"It is not clear what the evidence... was"*. On this, I would agree with Turner, but also point out how on that same page, the reader is being set up for a powerful lie.

Mixed with Turner's misleading sentences scouring for tangible dirt beyond Silvert inexcusably seducing a girl child, or Duvall's heavy-handed child therapy, he throws in an outrageously suggestive mention of the *Rene Guyon Society,*(!) *a pedophile group to which neither Reich nor any of his associates or trainees had any connections to or approval for whatsoever.* Turner could not end his discussion on the misbehavior of two of Reich's associates without implying, through introduction of irrelevant material, that *maybe* Reich *might have* advocated pedophilia!? *No, he did not.*

Turner's slander about the Rene Guyon Society is made all the worse when one considers how Reich's own *International Journal of Sex-Economy & Orgone Research* carried an article in 1944, which presented a highly critical review of Guyon, authored by Reich's student Alexander Lowen[72] (who later went on to found *Bioenergetics therapy*). It appears, Turner did his "research" by superficially leafing through Reich's journals, found the critical review of Guyon, but then possibly deliberately *reversed the meaning of it* through defamatory suggestive hints and implications.

This "dropping of hints" with false comparisons to perverse actions by unrelated others is exactly what Turner previously did while discussing Reich's therapy methods, by inappropriately and unjustifiably referencing unrelated and out-of-context "competing schools of masturbation therapy". By Turner's slanderous innuendos and inserted smutty suggestions, the uneducated reader is led to believe Reich not only "masturbated patients", but maybe also tolerated pedophiles within his circle. *There's not a shred of evidence for either of those smears,* but it is done in such a manner that Turner can claim, technically, he "didn't say that". But the implications are absolutely clear, as there is otherwise no reason for him to mention such things in his book. And again, I must draw attention to how Turner never, to my knowledge, criticizes Kinsey for his documented advocacy of pedophilia, nor his using a serial child rapist "Mr. X" as a source of "data".

No citations, references or documents are given to support the malicious suggestions Turner aims at Reich. Turner then mentions the departure of Dr. Allan Cott from Reich's organization, claiming that Cott left orgonomy because of something improper going on. However, Cott's departure was detailed by Lois Wyvell, Reich's personal secretary during those years, who wrote a series of memoirs in a self-published magazine *Offshoots of Orgonomy*.[73] In the Autumn 1983 issue, she recounted that Cott became frightened by the FDA investigations and related attacks in the press, as triggered years earlier by the Brady smear articles. Cott departed orgonomy in March 1952, by Wyvell's account, which said nothing about Cott making accusations against Reich or other orgonomic therapists, sexual or otherwise.[73]

Turner ends this ugly section of his book with a secondhand report about what Reich's angry ex-wife psychoanalyst Annie Reich-Rubinstein purportedly said back in the 1930s, as remembered in the late 2000s by their now 70+ years old daughter, psychoanalyst and father-hater Lore Reich. She was only a small child at the time and grew up with poison being poured into her ears by her mother. Turner quotes Lore Reich where she sounds to be just blabbering away (p.323-324) in an angry rant, without any apparent thought to fact or truth. He references her unsettled feelings about family photos taken on the beach, where few or no clothes were worn. Some of those photos were reprinted in the book by Ellen Siersted *Wilhelm Reich in Denmark* [74] and they are quite innocent. One photo shows Reich with his two daughters Eva and Lore, who are sitting on a blanket with the father Reich sitting behind. The two girls are nude, while Reich wears swimming shorts. Little Lore Reich is cringing in the photo, but not Eva nor her father.

Such is what Turner stoops to reporting about. Pathetic detailing of gossip and innuendo, and trying to turn innocent things, such as photos of a family on the beach, into something dark and dirty.

Some years ago, as detailed in the Appendix document, I exposed this problem with the "skeptics", that *many are open public advocates of the most extreme sexual perversity, including xxx-porn, sadomasochism, bestiality and pedophilia.* Those troubled individuals never had Reichian therapy with anyone, yet they nevertheless hate Reich with a severe intensity, *probably because Reich, more than any other modern behavior scientist, exposed their pathology on such matters via serious clinical and social analysis.* Reich also was no "Bible-Christian" who could be more easily dismissed by the atheist leftists who typically inhabit the "skeptic's clubs".

This entire section of the "Orgasmatron" book is nothing but a pile-on of outrageous slander, salacious innuendo and deception, given how *protection of children from adult seduction* was an explicit point in Reich's clinical and social work, starting with his SexPol platform[41] of 1932, and maintained over his entire life. One can review Reich's entire body of printed work, in many books and research journal articles, search the entire *Bibliography on Orgonomy,*[75] and one will not find even one sentence of support for the pedophiles, anywhere. Instead, one finds a constant theme of protecting infants and children from harm and trauma of all sorts, including and especially from sexual trauma. Reich was one of the primary psychoanalysts who argued that neurosis was founded upon *real traumas, including sexual traumas,* and not upon imagined "fantasy or wish" as per the usual psychoanalytical excuses. And that's partly why the psychoanalysts came to hate him, and to spread malicious rumors about his "going insane", which Turner so casually and maliciously repeats.

Reich's entire body of work gives much attention to protecting and rescuing children from bad life situations, to ameliorate whatever damage was done to them in the hospitals via heavy-handed obstetrical interventions, or by contactless and cruel caretakers, teachers and parents. Children are generally referred to orgonomic therapists because they are deeply suffering, being frozen emotionally, contactless or acting out in a dangerous manner. Frequently this can be ameliorated by simple measures, recognizing for example that the child is being subjected to parent-cruelty, teacher-cruelty or bullying at kindergarten or in grade school, and to simply protect or remove them from those situations. At other times, it is helpful or necessary for the child to mobilize their breathing and release frozen emotions. But the therapy Reich developed had no harsh cruelty advocated or built into it. Reich firstly wanted children to feel comfortable and unthreatened. When necessary, he got kids to cry by tickling them, or by playing with them as in one reported case by getting down on the floor on his hands and knees, and telling the child "Let's play doggy", barking and scampering around, panting like a dog with his tongue hanging out, and the child then delightfully copied him in play, started breathing deeper, and then out came the crying. He'd then comfort the child. It was playful, emotionally touching, not threatening, and my descriptions are typical of the many reports on Reich's sensitive and gentle approach to children.

His daughter Eva Reich built upon her father's example, developing her own soft-touch therapeutic methods that could trigger similar deep breathing and "stuck" repressed crying in distressed infants and children. Her gentle *Butterfly Baby-Massage* methods involve no pain or threats whatsoever, instead applying soft manual touch-pressure less or equal to that of taking a pulse, on specific key emotional-release points.[67] Nothing in Reich's journals nor in the later orgonomic journals published by his surviving trainees reveals any sort of hard abusive cruelty to children. Just the opposite. If there was child cruelty written up in any of Reich's publications, or in those of his associates, Turner would have pounced on it, and happily quoted it. But he does not, and cannot because such does not exist. So it becomes quite outrageous to see the truth repeatedly turned upon its head in the extrapolation of these very few examples of possible abusive conduct by one or two of Reich's associates, mostly after his death, to all of Reich's work and research legacy, if not also to Reich personally.

Aurora Karrer's Confusions, and Reich's Last Years

One final example of Turner's defamation of Reich's character comes from the confused and contradictory personal archive of Reich's last lover and common-law wife, Aurora Karrer. Reich had a stormy relationship with Ms. Karrer starting in the Summer of 1955 and lasting until March 1957, when he was incarcerated in Lewisburg Penitentiary. He died on Nov. 3rd of that same year. Mrs. Wilhelm Reich, as she frequently identified herself, died around the turn of the millennium, but she left an archive of letters and other documents at the *National Institutes of Health* (NIH) which cover the approximate 2 years of their relationship, including Reich's period of incarceration and some years afterward.[76]

Turner consulted the Karrer Archive, and some of his most damning condemnations of Reich emerge in that part of his "Orgasmatron" book, using Karrer-Reich as a reference. But he provides only cherry-picked material from that Archive, which I also examined for an independent assessment. I'll present my own view here, noting that I have not fully reviewed Reich's own personal notes and recollections on this matter for the years in question. In fairness to Reich, his side of the events must ultimately be examined. Those materials may eventually be published, and if so, will hopefully shed additional light on his relationship with Aurora Karrer, and on the events of that period.

In the early letters found in the Karrer Archive, Reich and Karrer expressed the most tender and deep love for each other. As the legal process ground on against Reich in his struggle with the FDA and US Courts, as the legal appeals were exhausted and imprisonment seemed likely, Reich understandably appeared to collapse into agony over the situation. By some accounts he occasionally drank heavily, and could be fully unpleasant at those times. Reich's feelings about the personal betrayals, sadness and aloneness, are apparent in his writings and photographs from the period, and he also clearly missed his children and former wife Ilse Ollendorff, who had moved away shortly before he met Karrer. And Karrer knew this, but stayed with Reich out of her own love and volition. However, Karrer's many written notes reveal her feelings at the time, and by my impression also suggest a growing resentment towards Reich, particularly regarding his remaining feelings for Ollendorff.

Ms. Karrer was 33 years old at the time of Reich's death at 60, but from what I can determine, she had no other man in her life afterwards. She left behind extensive and sometimes day-by-day diary-like reports, written on note pads or chaotically on whatever paper was handy, recounting her whereabouts and activities. The early entries reveal her love for Reich, and she outlined the start of a *Wilhelm Reich Biography* while living with him. The contents of this early outline, dated 22 July 1956, is a work of reasoned accuracy with appreciation for Reich's scientific work. She listed philosophers, musicians and artists whom Reich admired, and others he disliked, and made note of his favorite writings, his thinking on Einstein, Kepler, Darwin, Marx, and so on. It was an outline for a serious bit of work, but also one which apparently never even got started.

Following Reich's death in prison, again by my reading of her materials, she appeared stunned, emotionally lost, and increasingly resentful towards Reich. She had given up a medical technician's career at the NIH to come and live with Reich and help him in his work – except that his work was seriously interrupted at that time by the FDA legal prosecution. In a little more than a year, he was headed for prison. After his death, his oldest daughter Eva Reich stumbled in her role as Trustee of the Reich Estate, and Karrer made efforts to gain the Trustee position for herself, trying also to affirm her legal status as "Mrs. Reich". Those efforts were unsuccessful, and when the court-appointed Trustee, Mary Boyd Higgins, took legal action against her to acquire Reich's archive materials in her possession, she appears to have lost all compassion for her former love.

Karrer's archive subsequently reveals a woman preoccupied with her dead husband's past loves, and she drafted a new book outline with the rather sarcastic title *"The Genius: Personal Life and Loves of Wilhelm Reich"*. This new book idea was not organized or written up in any manner, but did have an outline of possible contents, revealing an unhealthy preoccupation with all the women whom her former and now deceased common-law husband had previously slept with and loved. Overall it shows a bitter sexual jealousy wherein her own angers and resentments periodically pour forth in the same diary-notes. Ultimately, she wrote lists of remembered or reconstructed events which at one moment are loving, appreciative and kindly towards Reich, but at other moments become terribly venomous, expressing extreme hatred. Very little of this material could be verified, but the character and tone of it suggest a woman in deep emotional grief and agony over her lost love, mixed with deep resentment and hate. On some of her older diary materials complimentary to Reich and written when Reich was still alive, she added marginal notations in 1970 or later which are completely without empathy for him, and also revealed memory confusions. The 1969 publication of Ilse Ollendorff's *Wilhelm Reich, A Personal Biography*[77] may have contributed to this reaction.

The Death of Troll

The death of Reich's dog, Troll, provides an example of Karrer's confusions and emotional turmoil, as she gave three different and very divergent accounts of it. In the earliest report, dated at 5 September 1955, she wrote: *"Troll... found by WR at the observatory in pain with broken right leg and left paw swollen."* Another report written sometime afterwards stated: *"About September 1, 1955 Troll was found nearly dead with broken right shoulder and broken left paw, and inner bleeding near to the observatory. He was hit by a bar or something heavy twice we believe. He is in the hospital under Eva's care."* Finally, some years after Reich's death, she wrote: *"...he [Reich] knew he'd beaten his dog Troll with an iron bar..."* [76] This last report came from some period after 1960, but exactly when cannot be determined. It could be as late as her angry 1970's entries. However, Karrer's later reports are fully contradicted by other written accounts of the dog's injury. Reich's ex-wife Ilse Ollendorff made note of it in her biography of Reich:

> *"Reich had been given a beautiful German shepherd dog in March 1953; a year or so later the dog was found one day with a badly broken hindleg – which may have been caused by an encounter with a car."* [77]

This quote from Ollendorff indicates Karrer was confusing things, that the dog was hit by a *car* and not a "bar". The Ollendorff account is further supported by documents in the Wilhelm Reich Archives at Harvard Countway Library, from early September 1955. One is a handwritten report by Eva Reich to the Sheriff in Farmington Maine, detailing the circumstances where: *"On 9/5/55 in the morning, Dr. Wilhelm Reich... found his 1-1/2 year old German Shepard watchdog unconscious, in critical condition, lying in a thicket about 50 feet from the entrance to the Orgone Energy Observatory at Orgonon, Rangely Maine. Mr. Alton Collins, Constable, was called as witness."* [78] Her report went on to detail the dog's multiple internal and external injuries, including broken bones and internal bleeding, as confirmed by a local veterinarian who made x-rays. The Reich Archive file also contained a separate report by the veterinarian, who wrote: *"In my opinion this dog was hit by some heavy object such as a **car**. This would account for the* *multiple fractures and internal injuries."* [emphasis J.D.] An exam was also made for possible poisons, which proved negative.[78]

This event happened only a short time after some local pestilent characters in Rangeley Maine, where Reich's laboratory was located, had carried out a string of provocative acts.[79] In one case, a group of teenagers led by the local pharmacist in the Rangeley community, paraded outside Reich's laboratory driveway shouting *"Orgy, Orgy, Commie, Commie".* That Reich was opposed to pornography and such things as "orgies", and became increasingly anticommunist in sentiments, starting as early as c.1930, was a point of fact lost on those citizens. Eva Reich understandably noted in her report on Troll, *"It seems unlikely that this was only 'an accident' in view of a long series of suspicious occurrences at Orgonon in the past"*[76,79] These independent reports from Eva Reich and the veterinarian at the time of the dog's injury, and later from both Ollendorff and the still-loving Karrer, indicated the dog was severely injured from being accidentally or deliberately hit by a car.

But as one might predict, *Turner ignored the Ollendorff report completely, making no mention of it, nor of the first report by Karrer from her own Archive at NIH.* Nor did he mention the Reich Archive file on this incident, which is clearly identified in the Archive Index as "Report on injury to WR's dog Troll".[78] *Turner only referenced the most malicious and final account by the confused, agonizing, and very angry Aurora Karrer,* of Reich allegedly killing the dog himself!

No legitimate historian would recount such an incident by cherry-picking from only the singular one report which best serves his premeditated and, in this case, quite hostile conclusion, of Reich being some kind of cruel monster. But then, Turner is not an historian. He's a liberal arts pop-magazine writer, who writes creative imagery, like paintings on a canvas, to impress the literary in-crowd and his fellow yellow journalists.

For the record, Turner attempted to visit the Reich Archive at Harvard, but was refused entry when the Reich Trust directors overseeing the admission policies learned of Turner's 2004 article in *LRB* attacking and slandering Reich. The Trust directors also learned Turner would include "Orgasmatron" in the title of his new book – he told them the book had the benign tite of *Unfinished Symphony: Wilhelm Reich's Sexual Revolution*, in what appears as yet another deception. Turner gives a mealy-mouthed complaint about being denied Archive access in the "Orgasmatron"

Acknowledgements (p.509-510), but misrepresents the reasons, just as he misrepresented his book title when applying for entry into the Reich Archive, and has repeatedly misrepresented Reich's biography and life-work.

As to the charge that Reich was occasionally drinking heavily in these last months before his incarceration, they may or may not be true, or could be partly true or exaggerated. But even if we assume the reports are true, Turner's recoundings were prejudicially hostile.

Reich suffered from the emotional shocks of being hunted down by the Nazis and Communists in Europe, having to constantly uproot his home and laboratory, suffering under slanders hurled against him in the Scandinavian press, then more slanders from the American press, the nasty rumors of the European and American psychoanalysts and psychiatrists, the FDA fraud and persecution, the book burning and public ruination of his name and research findings, leading to social isolation and eventually two heart attacks, the second one fatal while in prison. He felt alone and betrayed by nearly everyone, and had good facts to show there was *a communist conspiracy against him,* standing behind the FDA "investigation", though nearly nobody believed him on that either. He lost contact with his children. His pet dog was hit by a car, possibly deliberately by malicious people in his neighborhood. He also anticipated he would die in prison. All these issues piled up during his last months before being led off in handcuffs to Lewisburg Penitentiary.

So the man may have occasionally got drunk.

Who among us, under similar circumstances, would not possibly do the same? Other pioneering scientists of historical note, faced with public smears, lies and financial ruination only half as bad, simply committed suicide, such as biologist Paul Kammerer or mathematician Michel Gauquelin. Reich's heart attacks were probably, at least in part, related to the stress of the smear attacks and FDA investigation; he also witnessed America succumb to fascist book-burning, and a silent "going along" with it all by the professionals and media, or their actively joining in with it, which is exactly what he had witnessed in the lead up to the Third Reich. The same was true of the maligned scientist of *water-memory* fame, Jacques Benveniste, who suffered from a heart attack after being shamelessly assaulted over years by both "skeptic groups" and major "science" journals, dying a few years later. All these pioneering scientists essentially died from broken hearts.

In Defense of Wilhelm Reich

But Reich gets no sympathy from Turner, who appears to deliberately contort the facts to suit his impressionistic artistry: Reich is transmogrified into some kind of pornographic drunken womanizing Frankenstein monster, a child-abuser, patient-masturbator and "orgasmatron-quack" with zero confirmation of his orgone energy findings by others in the scientific or medical community, who also beat his friendly dog to death with an iron bar.

Thus is Turner's maliciously distorted portrait of Wilhelm Reich.

Summary

Historical revisionism and lies of omission saturate Christopher Turner's "Orgasmatron" book, while the essential facts of Wilhelm Reich's life and work are censored out or convoluted. Every ugly rumor or nasty accusation which has ever been remembered or written against Reich is dutifully dredged up from the bottom of the cesspool, reported in detail and magnified so as to lead the uninformed reader towards the worst possible conclusions. By contrast, nothing which validates Reich, or defends him, or which would overturn Turner's central thesis of "Reich the quack" or "Reich the pervert," is ever brought into discussion. Key citations in Turner's work do not check out; questionable materials are cherry-picked and included for derogatory content only; uncomplimentary factual material from a rare isolated instance gets generalized as a common or frequent occurrence; inappropriate comparisons of Reich are made to outrageous sexual criminals or groups to which he had not the slightest connections; and key facts are frequently twisted into their opposite. Which only goes to show what Goebbels and the Soviets knew and perfected in the mid-20th Century, that one can get away with very Big Lies by mixing them up with carefully selected half-truths and lies-of-omission.

Turner also frequently takes the tone of *Grand Inquisitor,* expressing feigned sympathy for the poor Reich tied to a stake, even as he pushes a literary burning torch into his face. As noted in several instances, some of the worst salacious slander came from one of Reich's ex-wives. How many reading this would like to be defined forever by their worst enemies, by debauched people or groups they don't associate with or even know, by the actions of people generations later, or by disappointed ex-lovers or angry former spouses?

Turner's book thereby creates a *Grand Theater of the Absurd*, a scripted play by a master illusionist, towards the pretense that there wasn't anything of merit in Reich's entire body of work. And that clears the decks for his *auto-da-fe* of Reich's personal and professional life. In the fictional artistry of Turner's world, Reich's orgone accumulator becomes Woody Allen's orgasmatron; Reich's clinical terms on the *orgasm reflex* and the *full genital heterosexual orgasm* are misrepresented as being either unattainable "sour grapes" for besmirching, or something jokey, found only inside the "orgasm box". Reich's emotion-release orgone therapy is redefined as "masturbation therapy", his criticism of pornography and frustrated neurotic hypersexual behavior is convoluted into a support for neurotic pornography and hypersexuality; his efforts to decriminalize homosexuality (while still considering it a neurosis) and his opposition to pedophilia (to protect children from adult seducers), his support for women's rights and human freedoms generally, all of it is turned around and upside-down. "The Life and Work of Wilhelm Reich" by Christopher Turner is more reminiscent of Dante's *Inferno,* than of the real-world factual situation. Of course, Turner revises history with a crafty smile, getting white-gloved applause from "all the people who matter" in the literary highbrow salons of London and New York, for whom book-burning and imprisonment of scientists is a mere "curiosity".

And what about Reich's books, which Turner largely ignores? So it is not surprising also, that *the Turner book makes scant mention of the FDA's fraudulent instigation of US courts to order the banning and burning of Reich's books, and his imprisonment, for basically daring to expose the fraud and defend his discoveries.* These are irrelevancies to Turner and his friendly reviewers, whose articles are laden with gloating and schadenfreude. Is this what passes today within the mainstream left-politic for "serious investigation and critique" of Reich's research findings? As "serious historical biography"? Apparently so.

One must ask, why is it that more than 50 years after his death, so many of these same "intelligentsia" and mainstream "literary journalists" get so worked up about Reich's findings that they not only write large books where nearly every page throws filth upon his grave, but a major publisher once very friendly to Reich's work – Farrar, Straus and Giroux (FS&G) – then proceeds to endorse such trash in a publicity campaign with hardbound and softcover editions, significant advertising and major media reviews? Did the publisher even bother to submit Mr. Turner's work to serious outside reviewers? Or if so, were these reviewers selected from the cadres of known Reich-haters? Did the "skeptic clubbers" and their fellow travellers in leftist political or porn circles, have anything to do with this?

From private sources, I know FS&G refused to allow the directors of the Wilhelm Reich Museum and Trust to give the Turner book a proofreading and fact check. It also appears that *nobody of scholarly merit was allowed to see it or review it prior to publication.* The whole enterprise spat upon not just Reich's grave, but also upon the grave of Roger Straus, former head of FS&G, who heroically stood up to the bully-boy FDA in the 1960s and championed the republication of Reich's banned and burned books when no one else would consider it. FS&G once saved Reich's work from oblivion. Now, they seem to want to destroy it.

Following the 2011 publication of Turner's "Orgasmatron", various "book review" articles appeared in the print and online editions of major news organizations on both sides of the Atlantic, all repeating and even amplifying upon the slanders. A partial listing includes *The Daily Mail, Daily Telegraph, Financial Times, The Guardian, The Herald Scotland, The Independent, London Times, London Review of Books, New Humanist, New York Observer, New York Times, New York Times Book Review, New Yorker, San Francisco Chronicle, Slate.com, Sunday Telegraph, The Economist,* and *The Wall Street Journal.*[65]

A documentary film entitled *The Sex Researchers* appeared later that same year on UK's Channel 4 TV, in a defamatory misrepresentation of Reich, also based upon the Turner slanders. In 2012, the science journal *Nature* also published a libelous article about Reich, authored by an "arts journalist" who used as her source a New York City museum display, which in turn got its disinformation from Turner. Both of these latter two incidents will be discussed in the next Chapter, but I am informed there are even more public slanders on the way, relying upon Turner's book for their "facts". The ugly rumors take on a life of their own, and since the mainstream "journalists" take on the aura of little gods, treating all complaints like "petitions from the peasantry, for whom the King's Court has no time to waste", nearly nobody publishes retractions or corrections, even when presented with excellent contrary documentation.

Against this tidal wave of mainstream media sewage, efforts

were made by myself and other professionals to write letters of rebuttal to the various newspapers and magazines. These never got into their print editions, but sometimes were tolerated in the online internet editions. However, as is the case with internet commentary, the correctives written by serious-minded scholars were drowned out by those posting up screeds which alternatively vented rage at the fabricated straw-man "pornographer Reich", or expressed a fawning titillation or laughing schadenfreude, *about* the false "pornographer Reich".

This close coordination towards the utter destruction of Wilhelm Reich and his research legacy, on both sides of the Atlantic, indicates an *orchestrated effort* in socially destructive character assassination. Reich termed this kind of organized malicious scheming and plotting in efforts to destroy truth and life, the *Emotional Plague*.[80] The Emotional Plague characterized the Nazis, the Communists, and also crept into the USA on the backs of organized Big Medicine, primarily in the structure of the Food and Drug Administration, which basically does the bidding of the major pharmaceutical firms and the worst of the organized hospital MDs.

But the Emotional Plague is also alive and well within other social contexts. It thrives within modern Big Media "journalism" where smiley-faced writers and film producers tell whatever lies are needed to gain access for interviews or photo shoots, after which the truth is cut to pieces. It is seen wherever intellectual or theocratic social manipulators and other "do-gooders" are at work "cleansing the social landscape" of dissenters and heretics in science and the healing arts. It is especially at work wherever issues of love, sexuality or children are concerned, with a goal to warp and twist the social fabric and the very lives of people towards chaos and destruction. There is no other goal in mind, except destruction of what is healthy and alive, in favor of the sick and the deadly. A review of Reich's findings on this specific emotional-sexual disorder, the *Emotional Plague,* is recommended and necessary.[80]

This current effort by Turner is only the most recent blow against the facts about Wilhelm Reich's life and work. Hopefully, the younger generations will seek out the facts as found in Reich's original writings. My hope is, as the facts emerge, Turner and friends will find the mud they have thrown blowing back in their own faces, while Reich is given a properly honored position in the history of science and medicine.

In Defense of Wilhelm Reich

**Open Letter to Jonathan Galassi,
head of Farrar Straus & Giroux,
regarding the Turner book:**

12 January 2012

Jonathan Galassi
Farrar, Straus & Giroux
18 W. 18th Street
New York, N.Y. 10011

RE: Slander, Defamation and Numerous Lies of Omission
in the Turner book *Adventures in the Orgasmatron*

Dear Mr. Galassi,

I am a natural scientist and former university professor with over 40 years of research expertise in the fields of Earth, Atmospheric and Environmental Sciences. I've also been directly involved, over that same period of time, in numerous research projects making direct experimental investigations and evaluations of the scientific findings of the late Dr. Wilhelm Reich. I am author or editor of several books on the subject of experimental orgonomy, and have written perhaps 100 papers on the subject, many of which are in mainstream scientific journals or academic-press books. I've also done considerable archival historical work on the subject of Reich's life, and the history of the attacks against him in both Europe and the USA, and so know quite a lot about the subject. So it was with some alarming shock for me to read the terrible slander and defamation carried in this new "orgasmatron" book which your company has regrettably decided to publish.

I've known the excellent role of Farrar Straus & Giroux in getting the Reich books out into the public, following the regrettable actions of our US Food and Drug Administration, which as you know both banned and burned those books. But this title by Mr. Turner, my God, what can I say except that it is indefensible a-historical slander and defamation of the worst sort, something you can only get away with because the target of the abuse is dead.

Attached is a copy of my recently developed article: *"Critical Review of Christopher Turner's Adventures in the Orgasmatron: A Defense of Dr. Wilhelm Reich and his Orgonomic Science Against Renewed Sexual Slander and Defamation"*

This article will give you the specific details about the severe problems with this book, as well as an introductory overview of the published clinical and experimental evidence in favor of Reich's science. The on-line version at the following URL will provide active weblinks:

http://www.orgonelab.org/DeMeoReviewOfTurner.htm

The fact that NONE of the documentary material summarized in my *Critical Review* appears in the Turner book, which nevertheless goes out of its way to present in great detail every ugly rumor, gossip and slander which has ever been uttered against Dr. Reich and upon which Mr. Turner adds his own considerable contempt, is so astonishing I could not believe it at first. And today, I still can hardly believe a reputable publishing house like FS&G which obtained the Reich titles under the guidance of Roger Straus, should want to have anything to do with.

So I must ask:

1. Who in heavens name was responsible for fact-checking and critical oversight of this book? Did anyone with a scientific background bother to review it, and I mean someone with knowledge of the published experimental reports supporting Reich's findings? Or was it shopped out to agents of the malicious "skeptic's clubs" or to the FDA for an opinion?! Maybe you don't know, there are many academics who hate Reich's theories so intensely that they typically drop all pretenses of rational objectivity, and make it a point to deliberately censor out of discussion anything weighing in favor of Reich. I know this from having worked in the universities over many years.

2. This book is classified as non-fiction biography, and so has got international applause from all sorts of apparently-coordinated book-reviews — virtually all of which repeat the worst of the Turner

slanders, and none of which show any kind of independence of thought. This suggests a group of paid-off "reviewers", and maybe that's the way books are marketed these days. I find it reprehensible.

You MUST read my *Critical Review*, which is already on internet and may eventually go out as a small book, in which case I plan to include this letter, along with an account of whatever you might do in response to it. At this point I assume you are a decent and honorable man who has simply been misinformed, or even deliberately lied to, by your own staff. If so, then I hope you'll now show some courage to do the right thing. My suggestions are:

A- Reclassify the Turner book as fiction-satire and return all the awards.

B- Pull the book from resale and/or access via Ingram/Lightning, and fire the people who were responsible for its oversight and development. Find out how and why such a slander-filled screed could ever have gotten out the front door of FS&G.

C- Show some spine and opposition to the ugly social deterioration where apologetics for defamatory slander and book-burning can gain such a happy following.

This is more than just a smear against Reich, but also throws merde upon the grave of Roger Straus, who had the courage to stand up against the FDA bully-boys and book-burners. Today, with this ugly book, *FS&G has changed positions and now stands with the book-burners*, facilitating their apologetics!

Kind regards,
James DeMeo, PhD

Director of OBRL   www.orgonelab.org
Publications list: www.researchgate.net/profile/James_DeMeo

*Neither Mr. Galassi nor anyone else at Farrar,
Straus and Giroux responded to this letter.*

5. The "Orgasmatron" Aftermath:
Slanders from the UK Channel 4/
Wag-TV *Sex Researchers* Program,
from *Nature* magazine, and *Wikipedia*

"The worthiest people are the most injured by slander,
as is the best fruit which the birds have been pecking at."
Jonathan Swift

Following publication of the Christopher Turner book *Adventures in the Orgasmatron*,[1] the mainstream press took aim at Reich, exploding into multiple outbursts of slander and mud-throwing. I've already given a short list of some of the major newspapers which were involved, in various print and online editions. But there were several incidents during the aftermath which were particularly egregious, in that they also took aim at Reich's scientific legacy and professional supporters today.

The Case of UK Channel 4 / Wag-TV[2]

In January 2011, a few months before the Turner book was published, I received an e-mail from a representative of Wag-TV and Channel 4 in the UK, regarding production of a new TV series they were developing, entitled *The Sex Researchers*. They represented this program as a *serious documentary* on the lives and work of various scientific investigators into human sexuality, such as Freud, Kinsey, Masters & Johnson, Magnus Hirschfeld, Havelock Ellis, and of course, Wilhelm Reich. They asked me for an interview contribution to the Wilhelm Reich segment of their planned program, and stressed how seriously they viewed their undertaking. The assistant producer of Wag-TV, Ruth Mayer, wrote:

"It seems clear that Wilhelm Reich is of vital importance to our story. It appears that too-often, his complex and wide-ranging

work has been oversimplified in the press, leading to a lack of understanding of its true significance. We'd like to show how his discovery of the physical biological energy and orgone can bring about a new understanding of drought, cancer cells, warfare and aggression among many other things. We'd also like to discuss the therapeutic properties of the orgone accumulator.

I realise that Reich's sex research was actually just one element in a huge body of work. However, I feel that this series will be a good way to highlight his important contribution to the field. I feel strongly that we should interview an expert who is continuing Reich's legacy, and carrying out further research in the same vein. For this reason, we would love to come and film an interview with you at the Orgone Biophysical Research Lab, and learn about how his discoveries continue to help and inspire people."[3]

Thus was the first contact and invitation I received, to contribute to what was represented as an authentic and honest documentary, by film producers who wanted to do a factual and accurate job. What developed thereafter, however, was dramatically different, and ended in a terrible slander of Wilhelm Reich's research legacy, as well as a sullying of my own name and work.

The principals involved were Director and Producer Stephen Kemp, and Assistant Producers Vicky Booth and Ruth Mayer, of Wag-TV. Oversight of the full program was made by Executive Producer Martin Durkin of Channel 4. It is impossible for me to know if all these people planned to deceive and create a malicious slander from the very start, or if the original impulse for their project was honest, and only later was steered into a cesspool by the clever manipulations of key persons. While they all must share responsibility for the dirty deed that was committed, we do know from Reich how one concealed emotional plague character can worm their way into formerly productive and effective organizations, to wreak havoc even while cloaking themselves as "deeply concerned" or "helpful". What I know for certain is, they made a persuasive outreach and appeal which sounded fully sincere.

By reputation, I knew some of the prior work at Channel 4, which made a top-notch set of programs exposing other serious scientific controversies. As mentioned in the last Chapter section, Channel 4 had produced the program on *Kinsey's Paedophiles,*[4] exposing his perverse interests and indefensible conduct in recruiting pedophiles to supply him with "data" on childhood sexuality.

Channel 4 also made a set of programs exposing the large scientific flaws in the HIV hypothesis of AIDS, notably referencing the works of Prof. Peter Duesberg,§ a top retrovirologist working in the *Department of Cell Biology at U.C. Berkeley.* Their programs, *The AIDS Catch* and *AIDS and Africa,*[5] summarized the bad science, the political suppression of scientists and physicians who dissented from the "official truth" about HIV and AIDS, andthe criminal cover-up of toxic side effects from "AIDS medications", in one of the world's largest pharmaceutical firms. These flaws lay at the foundation of the "infectious HIV" edifice and the subsequent public propaganda and "HIV=AIDS=Death" hysteria. The programs debunked the deceptive mainstream claims about what was causing AIDS, and they are still available for online review.[5]

Channel 4 had also produced *The Great Global Warming Swindle,*[6] which likewise criticized establishment views and exposed political conniving in the promotion of the *Industrial-Age CO2 theory* of global warming. Their criticisms supported what I knew and had personally observed in my own weather research, which embraced the existence of natural climate cycles, notably a pre-Industrial warming since the cold depths of the *Little Ice Age* in c.1700.

My views and positions against both the Industrial-CO2 and infectious-HIV theories had developed over 40 years of practical experience as a working natural scientist, well before I ever saw those Channel 4 summary productions. In fact I am publicly known in these regards. In 1991 I wrote a positive book review for Michael Fumento's *The Myth of Heterosexual AIDS,*[7] and in 1993, a summary essay article, *"HIV is Not the Cause of AIDS",*[8] which got widespread replication and internet posting. I was also a signatory to the 1991 *Open Letter To the Scientific Community* as drafted by the original 12 members of *The Group for the Scientific Reappraisal of the HIV-AIDS Hypothesis,*[9] and which was submitted to *Nature, Science, The Lancet* and *The New England Journal of Medicine.* Only after much difficulty was it finally published in *Science.*[9] In 2008 I also added my name to three public scientific petitions, dissenting against the CO2 theory of global warming.[10]

---

§ Additional detail on Duesberg's critique of "infectious HIV" theory is found on p.189-192.

In Defense of Wilhelm Reich

I mention these things because professionals have a primary responsibility to stand up and speak out against socially-destructive trends in medicine and the sciences. In 1975, I had taken a similar position against nuclear power plants, signing a public petition when I was but a mere undergraduate student, for which I caught hell from angry professors who supported atomic power. Only today is that hazard recognized more fully, but it took the disasters of Three Mile Island (1979), Chernobyl (1986), and again at Fukushima (2011), for the general public to stand up against the scientific and political leadership and put an end to the uncritical government support being given to nuclear power. One can only imagine what disasters will be required to wake people up to the flaws in the CO2 theory of global warming, or to the flawed "infectious HIV" theory of AIDS. A new global chill like the Little Ice Age, which brings stormy disasters going against all the "politically correct" preparations for warming? Some new and dictated "AIDS Vaccine" which kills thousands of children?

My views have not changed on these important issues, in spite of dissenting voices being continually slandered, fired from jobs, and being censored out of public discussions, as "AIDS Deniers" and "Climate Deniers". The typically left-wing politicians, scientists and journalists whose intolerance has led to this situation deliberately use the "denial" word to connect every critic of those flawed mainstream theories to the Jew-hating Holocaust deniers. That fact alone should alert honest people to the ulterior motivations of the CO2/HIV extremists.

In all of the above cases, Channel 4 took on very controversial subjects which politically- and economically-motivated power-cadres within establishment Big Science and Big Medicine have tried to cover up and sequester for years. Channel 4's programs presented the arguments and evidence from the scientific dissenters in a serious and dignified manner, and exposed weaknesses in the orthodox mainstream claims, as well as the censorial abuses which the dissenters had experienced.[11]

In their approach to me, representatives of Channel 4/ Wag-TV personally stated they knew about the on-going defamatory slander of Reich, knew about life-energy from different contexts (such as acupuncture and *subtle-energy* research), and wanted to help set the record straight. For these reasons, in addition to the friendly communications received from Wag-TV, I had logical reason to believe they and Channel 4 would make a similar strong and serious presentation on Wilhelm Reich's work, thereby to oppose decades of slander, defamation and censorship by the mainstream of media, medicine and scientism. Not once in 40 years of doing orgonomic research had such an opportunity developed. While I had some remaining unsettled feeling (which in retrospect, I should have listened to) I decided to proceed with helping them to produce, as they had expressed, *something good and decent.*

Over the next several months leading up to the interview, and afterwards also, I sent them various research papers and other published materials supporting Reich's original findings on human sexuality and on the orgone energy, including details on many of the newer studies from within peer-reviewed scientific forums, as already discussed in Chapter 3. These included my book *Saharasia,*[12] detailing the largest-ever cross-cultural study on human behavior, analyzing sexual and family life in over 1100 different world cultures. I also sent Channel 4 copies of my books detailing new evidence supporting Wilhelm Reich's orgone energy discoveries, such as the *Orgone Accumulator Handbook,*[13] and *Heretic's Notebook: Emotions, Protocells, Ether-Drift and Cosmic Life Energy.*[14] They were additionally provided with weblinks to my YouTube videos on Reich,[15] as well as several private PowerPoint presentations I had given to scientific conferences in recent years.

In my communications, as well as in the subsequent interviews, the slanderous and defamatory nature of Reich's critics on the issues of human sexuality were discussed. Reich was not any kind of "pornographic character", I emphasized, and was in fact quite modest, and had even been contemptuously ridiculed as "a prude" by modern promoters of the compulsively promiscuous "anything goes" sex-bath-house and "swinger" scenes.

I made efforts to summarize Reich's view of genitality and his sex-economic theory, and how that is so different from every other theory on human sexuality, including or especially the "raw discharge" ideas of Kinsey, for whom any and every kind of sexual stimulation was considered of equivalent personal and social health benefit. Reich's concepts of the full genital orgasm, versus the lesser "sexual climax", was also mentioned to the Channel 4 team, not just in the printed materials I provided, but later during the actual interviews which were made. Our correspondence leading up to the actual interview was professional and courteous, and I anticipated they would do an excellent job, sticking with the facts.

However, in this case I was utterly and completely wrong.

The Wag-TV crew spent a full day at my institute, the *Orgone Biophysical Research Laboratory* (OBRL) in mid-February 2011, where all the above matters were discussed and recorded on video. They also took video shots of my laboratory, the orgone accumulators, and the cloudbuster. While the interview itself went reasonably well, lasting over several hours and covering many different aspects of Reich's life and work, and of my work, it was in the editing where severe betrayals and back-stabbing occurred.

After the interviews were finished, I was informed that a copy of their film segment on Reich, which included my interview segments, would be sent to me in the mails on a DVD disk. This would be, they said, sometime just before or around the time it was broadcast on British TV. In fact, I never got anything from them.

Months of silence passed. In early July, friends in the UK informed me that the *Sex Researchers* series had recently been broadcast on UK television.[16] By their observation, which I confirmed in a subsequent internet download, it was *a first-class hatchet-job, making Reich out to be an insane quack and pornographic charlatan.* Worse for me, cuts from my interview were carefully taken out of context and spliced into the program to give the deliberately false impression that I somehow agreed with their misrepresentations of Reich. The sequence on Reich lasted only 6 minutes of a 45 minute program, but was a testament to the worst kind of defamatory slander one could imagine.

Other than showing a few parts of my interview and laboratory, their Reich segment used actors making fictional "reenactments", supposedly from his therapy room and laboratory, with a voice-over by the British actor and comedian, Robert Webb, whose words dripped with snide contempt, disguised as "humor".

After giving a very brief introduction, one of the early scenes in the program showed the Reich-actor masturbating himself while hooked up to a primitive voltmeter, standing with his back to the camera in a darkened bedroom with drawn shades. The narrator had this to say during those scenes:

NARRATOR: *Reich thought he had discovered a type of ...invigorating energy field ...fueled by the power of the orgasm.*

The "Orgasmatron" Aftermath

He called this energy orgone, and set about trying to measure the electrical discharge of his penis during masturbation. Reich ...encouraged people to have as many orgasms as possible. This would he felt lead to health and harmony throughout the world. [16]

This scene and words utterly misrepresented Reich's bioelectrical experiments on pleasure versus anxiety, as conducted with volunteer test subjects in laboratory rooms at the *University of Oslo Psychology Department*. Those experiments had been supported by Harald Schjelderup, physician and founder of the Department, who had assisted Reich's flight from Nazi Germany. Reich's experiments were conducted professionally with much sensitivity, entirely opposite of the fabricated scene of self-masturbation in a darkened bedroom. That work also took place well before Reich's discovery of the orgone energy, but no matter, the film just mixed up whatever was necessary to make dirty-minded jokey innuendos. I've already discussed Reich's bioelectric experiments in the prior Chapters.

In other scenes, the Reich-actor was shown in a therapy room, sitting on a chair near to where a naked young woman lay down on a bed. The Reich actor leaned towards her as the camera zoomed in on her face and shoulders, after which she writhed and moaned with pleasure. This scene deliberately misrepresented Reich's emotion-release *vegetotherapy*, or *orgone therapy*, as some kind of "masturbation therapy", which it most certainly never was.

NARRATOR: *Reich ...encourag[ed] his patients to undress and pressing hard on what he termed their body armor, until the patient climaxed and released wave upon wave of orgone. (naked woman on therapy couch sighs and squirms at this point) This became the chief goal of the Reichian therapy, each orgasm topping up the worlds supply of orgone.* [16]

The Reich-actor feigned to masturbate a naked woman actress to sexual climax while on the therapy couch. This particular slander came from Reich's angry ex-wife and psychoanalyst Annie Reich-Rubinstein, as discussed in the prior Chapter on the Turner book.

The program narrative also sarcastically ridiculed Reich's discovery of the orgone energy as "orgasm energy", falsely implied the orgone accumulator was used for masturbation, and invented other scenes which never happened in real life. These included a scene where the Reich-actor slumped down to the floor, and then rubbed his hands over his head and messed his hair, rolling his eyes around in feigned insanity.

> NARRATOR: *Although popular with some of his patients, for many of his colleagues, Reich had crossed the line. Questions were asked about his methods, and as rumors began to spread that he was mentally ill, he fled Europe for the American wilderness. Here he reinvented his career and began constructing special boxes to harness orgone energy, his famed orgone accumulators. ... The idea is that this global orgasm energy is somehow sucked into these boxes and then radiated from the metal walls towards the person sitting inside.*[16]

This last contempt-filled nonsense also claimed, falsely, that Reich fled to America due to sexual scandal. As previously discussed, *Reich fled to America to escape the Nazis and Communists whom he had publicly criticized, and who had put his name on their respective death-lists.*

The reader will also note the dark similarities throughout this narrative to the Turner "Orgasmatron" book.[1] I will expose more details about this shortly.

My interview segments were selectively edited and spliced into the program at specific places basically as props and scenes where some electronics or apparatus could be shown against the narrator's slanderous words. I hardly spoke, except when they could lift out a sentence or two to fabricate the false impression that I was in agreement with their smutty view of Reich. In so doing, the program was as defamatory against myself as against Reich.

Fortunately, the UK has strict laws against libel, defamation and character assassination. While Reich was dead and could not file a lawsuit, I could do so. And the outrageous nature of the slanders, their complete falsity, and the way the film crew had lied to me repeatedly about their desire to do an honest documentary in order to get my help and consent, left both Channel 4 and Wag-TV, and the various directors, producers and executives open to this possibility of legal actions.

I consulted with friends and lawyers in both the USA and UK, and steeled myself for what would be an expensive and lengthy legal battle. However, there were some preliminary steps which could possibly make legal action unnecessary. The UK has a government organization called *Ofcom*, which is empowered to hear complaints from the public about unfair treatment by broadcast media.[17] When a complaint is made to Ofcom about a specific program, they investigate the matter, gather evidence from the various parties, and render informed judgements.

I therefore made a formal complaint to Ofcom. In addition to informing on their defamatory treatment of Reich and myself, I also detailed the numerous lies they had committed, in spoken word and e-mails, about how they had wanted to do a serious and fact-based documentary, with apparent deceit and trickery to persuade me to grant an interview and open my laboratory to their cameras.

In early July 2011, I sent Ofcom a formal *"Complaint on Sexual Slander, Character Assassination and Defamation by Channel 4 / Wag-TV".*[18] This was followed some months later by a "Rebuttal" from Channel 4/Wag-TV,[19] which constituted mostly a plea of innocence and a denial about what they had done. However, it was illuminating on some important points about how and wherefrom they got their distorted ideas about Reich, as I will mention below. I thereafter provided Ofcom with a more detailed *"Response to ... the Channel 4 Response."* And so it went for the next 6 months.

My Complaint to Ofcom put an immediate halt to Channel 4/ Wag-TV plans to make a wider distribution of their *Sex Researchers* program in rebroadcasts and as a DVD production. Ofcom forced them to stop further distribution of their materials until a formal investigation and ruling would be made.

As the Ofcom procedures developed, it came out that Channel 4/ Wag-TV had indeed consulted with Christopher Turner and his "Orgasmatron" book to make their slanders against Wilhelm Reich. *The similarities between the dirty-minded distortions of Reich in the Sex-Researchers production and the slanders in the "Orgasmatron" book were therefore neither coincidental nor accidental.*

In their Ofcom Response,[19] Channel 4/Wag-TV several times referenced *"Reich biographer Christopher Turner"*, or *"Christopher Turner, who has written the latest, authoritative biography of Reich (Adventures in the Orgasmatron), based on numerous interviews with people connected to Reich and his own extensive research."* Turner's terribly slanderous book, filled with smutty distortions about Reich, was unjustifiably classified as an "authoritative biography" – via sheer chutzpah and indefensible declarations from his publisher, and *without any outside scholarly review or fact-checking* – after which others then cited his fabrications as a new "truth".

From these revelations, it appeared Channel 4/Wag-TV had been consulting with Turner and his poisonous book at the same time I was providing all the accurate information to them about Reich and his research findings. They chose, perhaps from the very start, to discard my documentation and instead to favor of Turner, because they could never produce such a lie-filled, dirty-minded docu-spectacle using me as a reference. My prior critique and rebuttal of the Turner book has also served as a corrective against the defamatory slanders by Channel 4/Wag-TV, and so do not need to be repeated here, except to make note of them.

In the end, *Ofcom upheld my complaint*, at least on the narrow grounds of "informed consent" and "fairness". Ofcom ruled I had been deceived into believing they were going to make an authentic and fact-based documentary, whereas the final program was some-thing entirely different. The Ofcom ruling carried a tacit under-standing that the final documentary was not honest in its presen-tation, as otherwise the issue of "fairness" (and ergo, "informed consent") had no merit. It was unfair because it was a defamatory slander against Reich, against myself, and against historical fact and truth. If my objections on points of fact had been untrue or frivolous, they would have ruled against my Complaint.

The Ofcom ruling is now a matter of public record as published in the *Ofcom Broadcast Bulletin*.[19]

In our last communications and following the Ofcom ruling, Channel 4 and Wag-TV informed me they would remove all of my interview sequences from their *Sex Researchers* program, for any rebroadcasts or DVD versions. However, I wrote to them with encouragements to *do the right thing and remove the defamatory smears against Wilhelm Reich as well*. It remains to be seen if this will be done.

The Case of *Nature* Magazine[20]

Nature magazine, considered one of the "top" science journals, has long been known to be heavily biased in favor of mainstream allopathic MD-hospital medicine and against natural healing meth-ods of all kinds. They also have a cozy relationship with the anti-science "skeptic clubbers". In one informal review I made in the 1990s, when I still had a subscription, *Nature* magazine had a full 50% of its pages devoted to full-color advertisements by pharmaceutical and biotechnology firms, or by the equipment suppliers to those industries. The articles published in its pages generally served to promote those same commercial interests, in the manner of a *trade journal*, as opposed to an authentic scientific one.

Under their prior editor John Maddox, who was closely allied with the CSICOP "skeptic clubs,"[21] *Nature* openly attacked or censored many scientists whose work threatened orthodox medicine, pharmaceutical interests, or the well-defended (and well-financed) fortresses of modern, but frequently dead-ended, mechanistic biochemical and astrophysical theory. Most notable here were the censorious attacks made against scientists Jacques Benveniste and Peter Duesberg.

Benveniste had demonstrated antibody reactions within *in-vitro* cell cultures exposed to an extremely dilute antigen, at below Avogadro's number. Biochemical theory predicts no possibility of cellular reactivity, but this was demonstrated nevertheless, thereby providing strong support for a non-chemical "water memory" mechanism at work within biology. And that gave direct support to the frequently ridiculed principles of homeopathic medicine.

Before they would consent to publish such a paper, *Nature's* Maddox demanded Benveniste's protocol and results be reproduced in a private testing laboratory, which was done, once more confirming the original antibody-reaction result. With much quibbling, *Nature* finally published Benveniste's article, but before doing so sent a "hit squad" of nonscientist skeptic-clubbers – including the one-time CSICOP darling and grey-beard magician James Randi (who separated from CSICOP shortly after public revelations of his soliciting sex from a teenage boy; this is relevant, see the Appendix) – into Benveniste's laboratory to "evaluate" his methods and procedures. Not surprisingly, the Nature "hit squad" had no serious interests to "investigate", but instead made a deliberate chaotic disturbance and later slandered the naive Benveniste in a separate article in *Nature* which accompanied the publication of his original findings. A *Nature* editorial tried to make a laughingstock of Benveniste, but in largest measure these antics – again, in the "best, top" science journal on Planet Earth – merely certified the scientism and unethical attitudes of the editors when confronted with truly

In Defense of Wilhelm Reich

novel and challenging research findings.[22]

Duesberg is a top retrovirus specialist who over the years has successfully made a strong scientific case against the dominant hypothesis of "infectious HIV" as the cause of acquired immune deficiency syndrome ("AIDS").[5,8,9] He and his associates had stymied a big push by pharmaceutical interests to steamroller through the claim that "HIV causes AIDS", and to thereby justify a raft-load of new and expensive "medicines" and vaccines. The "infectious HIV" critics had the facts on their side, however, and with Duesberg being one of the world's top retrovirus specialists, their Group constantly pointed out the serious flaws in experimental designs and claimed "proofs" by the many HIV advocates. Notable here was his charge that a typically prescribed "medicine" used to treat clinical AIDS symptoms, *Azidothymidine* (AZT), was in fact the cause of many AIDS deaths. AZT originally was an ineffective and never-patented experimental chemotherapy drug. It was resurrected from the pharmaceutical graveyard to treat AIDS patients, however, with deadly results, killing up to 75% of those under treatment by some studies.[5] And since the clinical symptoms of AZT poisoning are indistinguishable from clinical AIDS (AZT is a DNA-chain terminator which destroys the immune system), AZT poisoning was being blamed on an incoherent theory of "infectious HIV".

Immune stress and breakdown certainly is a problem for specific high-risk groups (i.e., drug-abusing antibiotic-dependent homosexual men, IV drug injectors, hemophiliacs, healthy people uncritically following "doctor's orders" to ingest toxic pharmacy drugs), but no truly infectious epidemic has ever broken out from those risk-groups. Other major problems exist, such as the many false positive reactions on flawed "AIDS tests", their cross-reactions with normal blood antibodies in healthy people, and the illness incurred when healthy people begin to ingest the extraordinarily toxic "medicines". Overall, the AIDS malady took on the specter of a classical *iatrogenic disorder, created by the physicians' own misapplication of deeply flawed theory, inaccurate tests and deadly medications.*

The point is, the arguments *against* infectious-HIV theory are quite serious, scientifically coherent and valid, but nevertheless are suppressed by those furiously addicted to their own Grand Theory, or unethically pushing vested financial interests and chasing Nobel Prizes. And in this they get the full support of the mainstream press, the same press being identified here as carrying forward a massive

190

campaign of slander and disinformation against Wilhelm Reich. And certainly not unrelated to Reich's larger findings on widespread sexual confusions and fear or hatred of genitality, HIV theory is advanced by various finger-wagging journalists and medical moralists. They not only preach the gospel of "sex can kill", and "everyone is at risk" to heterosexuals and to schoolchildren, but are sure to include all kinds of "gay" propaganda which censors out discussion on the highly unsanitary, immunity-destroying homosexual bathhouse practices, or the foreign blood proteins which always sickened people injecting street drugs or receiving multiple blood transfusions. HIV promoters today also advocate ancient tribal genital mutilations – circumcision – as an "AIDS preventative", warn mothers away from natural breast-feeding, and promote other superstitions, all to the glory and finances of the "heroic" hospital-MD white-coats and pharmaceutical houses. This is done, even as they and their allied journalists sometimes bitterly complain that "AIDS Deniers" aren't being imprisoned, with their publications burnt – as happened to Reich!

A group of physicians, professors and a few Nobel Prize scientists joined with Duesberg to expose this public health nightmare. But the AIDS juggernaut could not wait any longer. *Nature* aggressively ended the scientific debates with an ex-cathedra declaration from editor Maddox that infectious HIV was the "cause of AIDS", so everyone should just shut-up about it. No further discussion or critique was tolerated from Duesberg or his associates.[23]

The abuse heaped upon both Benveniste and Duesberg spread into other corners of medicine and science, typically including unethical "skeptic" type attacks replete with "dirty tricks". Mental gymnastics were applied to justify the censoring of further discussion on those issues. Both men were top-level scientists of good reputation. Benveniste was a senior scientist at the French INSERM laboratory. Duesberg was a professor of Cell Biology at UC Berkeley, a member of the National Academy of Sciences with a 7-year *Outstanding Investigator Grant*, and the winner of many awards for his pioneering work on retroviruses. A full discussion of their findings and the unethical attacks directed against them is beyond the scope of this book, but can be found elsewhere.[5,7,9,11,22,23] I mention their cases, as they have become textbook examples of how modern medicine has descended into *Medieval Scientism*, and how pharmaceutical and other interests work behind the scenes to

In Defense of Wilhelm Reich

silence the opposition and get their way, no matter what are the empirically determined facts.

The attacks against Benveniste and Duesberg, aiming to effectively destroy their work and professional standing and to subvert the scientific reception process, were carried forward by the same sort of "journalists" and "skeptic-club" brigades currently under discussion in the case of Wilhelm Reich. The whole of modern scientific thinking was forcibly steered away from all their findings, away from their scientific and medical discoveries, and away from their critiques of mainstream errors, to the severe detriment of public health.

With that background, it was not too surprising when in August 2012 *Nature* allowed an arts journalist, Josie Glausiusz, to write a full-page gossip column, wherein several paragraphs slandered Wilhelm Reich.[20] In her article, Glausiusz merrily chatted up her visit to the *New Museum* in New York City, where she reviewed a display entitled *Ghosts in the Machine,* which included defamation against Reich. Glausiusz' *Nature* article, being an "arts" item, carried no references, so at first it was unclear where she had got her misinformation. She maliciously repeated the Brady-Gardner slanders from the 1940s, falsely accusing Reich of offering the orgone accumulator as a method for "curing cancer", and as a sexual appliance which could "increase orgastic potency". Glausiusz also nastily, and wrongly implied that Alan Turing – a member of the British Bletchley Park code-breaking team which cracked the Nazi "Enigma Machine" ciphers, advancing the Allied cause during World War II – was forced into Reich's form of body-oriented psychotherapy, "to cure his homosexuality", after which he possibly committed suicide in 1954. The article also claimed that Albert Einstein had examined Reich's orgone accumulator and had "pronounced it a dud". All these malicious fabrications were condensed into two miserable paragraphs, leaving the uninformed reader with a most terrible false impression. As so:

> *"Another misconceived machine in the show, created by psychoanalyst Wilhelm Reich in 1940, is the Orgone Energy Accumulator, a wooden cubicle lined with sheet metal and steel wool. A follower of Freud, Reich believed that a concentration of "orgone", which he conceived of as a primordial form of energy, would increase "orgastic potency". This would, in turn,*

relieve pent-up psychic and physical tensions, promoting general health and vitality, and even playing a part in curing cancer. Reich offered the box for testing to Albert Einstein, who pronounced it a dud.

Among those treated with orgone therapy was British computer-science pioneer Alan Turing, who was forced into treatment to "cure" his homosexuality. His proposed Turing Machine would simulate the behaviour of any other machine, including itself. Two years after being prosecuted for gross indecency in 1952, Turing committed suicide, possibly by eating a cyanide-laced apple.

Henrik Oleson's Some Illustrations to the Life of Alan Turing *(2008) appears in the same room as Reich's accumulator."* [20]

An investigation clarified how Glausiusz got her "facts" wrong, notably from a collusion between her and the New Museum curators, who had personally developed the *Ghosts* display. Both the Museum curators, and Glausiusz, it was discovered, had also relied upon Christopher Turner of "orgasmatron" fame in the writing of the *Nature* hit-piece. More on this in a moment.

After the Glausiusz article appeared, I began work on a rebuttal article for submission to *Nature*, titled *"In Defense of Wilhelm Reich"*. My rebuttal corrected the Glausiusz errors and presented a summary discussion on new published experiments affirming and verifying Reich's original findings.

For example, as detailed in the prior Chapters, Reich never offered the orgone accumulator as "a cure for cancer" nor to "increase orgastic potency". Reich also had no British trainees in his therapy methods at that early date, and so could not have been involved even remotely in whatever therapeutic treatment Alan Turing was subjected to. In fact Turing was given hormone injections by court orders when his pedo-homosexuality became publicly known. The 40-year old Turing, supposedly under tight wartime security restrictions forbidding him to meet with or even speak to strangers, had basically anonymous sexual relations with an adolescent teenage male. His apartment was subsequently burglarized, and he reported it to the police, pointing to the boy as a suspect. The whole matter then erupted into a public newspaper scandal.[24]

As also detailed in Chapter 4, while Reich considered homosexuality to be a neurosis, and viewed pedophilia as a crime against which youth had to be protected, he was sympathetic to the plight of homosexuals and openly advocated for decriminalization when children were not involved.[25] His therapy methods did not include the use of psychopharmaceutical drugs, in any case.

As to Reich's 1941 meeting with Einstein, this also is covered in prior Chapters. After their initial correspondence, Einstein invited Reich to his Princeton home where they had a serious discussion lasting 4-5 hours. After observing Reich's demonstration of anomalous temperature increases within a special orgone accumulator, Einstein spent several weeks making independent tests, declaring in a letter he had *confirmed* the thermal anomaly.[26]

Everything journalist Glausiusz wrote about Reich in *Nature* was either a misleading half-truth or an outright falsehood. Her language additionally left the reader with the malicious impression of Reich as a sleazy crackpot whose therapy methods were potentially deadly. While the overall *quantity* of slander in the *Nature* article was small by comparison to other published smears, to have such material appear in the pages of *Nature* magazine suggested a more deadly situation, in the further building of historical revisionism at a supposedly "high scientific level".

With some rational outrage, I thereafter organized a public call within private professional circles, to *Stand Up for Wilhelm Reich*, and to gain critical feedback and additional signatures for the rebuttal article, which would then be presented to *Nature* for publication. I was most pleased at the positive response. A total of 23 additional professional scientists and physicians with the MD or PhD degree, as well as a few MSc and PhD candidates, signed on to the article as a public protest against the on-going slanders of Wilhelm Reich and his orgonomic science. The revised rebuttal article, with 24 signatories who gave it their review and proof-check, was then submitted to *Nature* for publication.

Unfortunately but perhaps predictably, *Nature* rejected the article within 24 hours, also deleting it from their online submission system, thereby making appeals of the rejection impossible.

A shorter *Letter to the Editor* was then composed and submitted, with the same list of 24 signatories. This did gain some traction, especially when I informed *Nature* we would publish the longer article in another peer-reviewed scientific journal. Two months of back-and-forth e-mails then transpired. *Nature* editors finally refused to publish the shorter letter, but they did agree to print their own "Correction" to a few of the falsehoods in the Glausiusz article. However they refused to retract any of the sexual slanders. Here is what they printed in their 8 November 2012 issue:

> *"Correction: A review of 'Ghosts in the Machine' (J. Glausiusz Nature 488, 279; 2012) reiterated incorrect information provided by the exhibition that Alan Turing underwent orgone therapy to 'cure' his homosexuality; he was in fact treated with hormones. Furthermore, Einstein did not actually pronounce the orgone energy accumulator "a dud" (for details, see his letter of 7 February 1941 in W. Reich The Einstein Affair Orgone Institute Press; 1953)."* [27]

This public retraction by *Nature,* while incomplete, was something of a small victory, a "first" in the history of Orgonomy, that a major science journal (or any other major media) would reconsider and retract a previously published derogatory attack against Wilhelm Reich and print a correction. However, important open questions remained. How could it be, for example, that *Nature* would permit such an amateurish and gossipy hit-piece as Glausiusz's to be published in the first case? How does sexual slander obtained by a city-beat arts journalist, repeating "stuff she heard" at an art museum in New York City, get published in a major British science journal without any fact-checking?

Shortly after the *Nature* correction appeared, new information was learned about why the Glausiusz article repeated sexual defamation against Reich, and why *Nature* refused to retract the worst of those slanders, even while correcting the error on the issue of Turing's suicide and what Einstein didn't say.

The original "art display" at the New Museum was prepared by the team of Massimilano Gioni and Gary Carrion-Murayari, who also wrote a book about it with the same title, *Ghosts in the Machine.*[28] Therein the source of the defamation was revealed. On page 7 they stated: *"A number of individuals worked closely with us to secure individual loans for the show. We extend our thanks..."* This was followed by a name list of around 100 persons, including one *Christopher Turner.*

Starting on page 127 of *Ghosts* one also finds a photostat reproduction of one of Turner's articles slandering Reich, *"Thinking*

In Defense of Wilhelm Reich

Inside the Box". This ugly little screed first appeared a year earlier in Turner's own *Cabinet* magazine,[29] and like his "Orgasmatron" book, it deliberately deceived and made Reich out to be a sex-obsessed quack. The article, reproduced fully in the *Ghosts* book, and in an under-glass display at the New Museum, included several photos, specifically one from the 1950s where an FDA agent posed with several of Reich's devices, including one stuck on his head, deliberately making a clown of himself. Other photos were shown of the fictional "orgasmatron" from the Woody Allen movie, *Sleeper*, placed suggestively next to a picture of Reich's physician daughter Eva Reich, sitting in an orgone accumulator.

That *Cabinet* article, faithfully reproduced in the *Ghosts* book and in the New Museum exhibit, was identical to one which Turner wrote for *The Guardian* newspaper of London in July 2011, but where the only photo was a group of nude women dancing in a circle out in a field.[30]

Such juxtaposition of false-comparison imagery has become something of a hallmark for Turner and his supportive friends in this modern form of *slander journalism*, as in the side-by-side display of the orgone accumulator and "orgasmatron", or the positioning of the name of Wilhelm Reich next to photos of naked exhibitionism, or implied medical quackery. Or as *Time* magazine once did, placing Reich's photo next to those of Stalin's Lysenko and Hitler's Mengele.

The New Museum *Ghosts* display and book thereby spread old slanders against Reich in new "artsy" packaging, courtesy of Turner. It also fabricated a connection between Reich and Alan Turing's suicide, which further confirmed how Glausiusz got her disinformation. Page 15 of *Ghosts* boldly but wrongly declared:

> *"After being identified as a homosexual, Turing was arrested and forced to undergo a variety of treatments in order to allegedly cure what were believed to be deviant and uncontrollable impulses. These treatments, which included Reichian Orgone therapy and the administration of female hormones, were an attack on the legitimacy and autonomy of Turing's body and likely contributed to his depression and eventual suicide."*[28]

At some point in November 2012, after *Nature* published a

limited correction to the Glausiusz's article, and righteous outrage was increasing among scholars and scientists who knew the truth about Reich's work, everyone involved including Gioni and Carrion-Murayari, came under criticism for what they had written. An on-line article about the "Ghosts" display, titled simply *"The Body Is a Machine"* and available from the New Museum website, did add a short corrective note of its own:

> *"This essay previously stated that Turing underwent Reichian Orgone therapy. The therapy that Turing underwent was known at the time as "organo-hormone" therapy and was not connected to "Orgone therapy." We apologize for this error."*[31]

This corrective did not, unfortunately, extend sufficiently for Gioni and Carrion-Murayari to distance themselves from the full Turner article, *"Thinking Inside the Box"*, which contained numerous falsehoods and slanders throughout, condensing much of what appeared within his "Orgasmatron" book. Hopefully they will do so in the future. And while they are at it, they should delete the photo suggesting boy-pedophilia on page 13 of their book. That photo, the first to appear in their book, is of the sweet-faced 5-year old Alan Turing, with a long screw laid on top of the photo, suggestive of "screw-boy".[28] Even if this was an unconscious *faux-pas*, it nevertheless screams out for removal. (I am always amazed at how those who go out of their way to assault Reich with sexual slander reveal, in one way or another, their own blind spots and unresolved personal issues.)

In December 2012, the full rebuttal article *"In Defense of Wilhelm Reich: An Open Response to Nature and the Scientific/Medical Community"* was published, with 24 professional co-authors, in the peer-reviewed science journal *Water*.[32]

Wikipedia Lies About Wilhelm Reich and Orgonomy

Wikipedia website entries are popular, but can be altered by virtually any anonymous person who desires to scribble some unsubstantiated personal opinion, gossip or deliberate deceit into their pages, or to erase legitimate and accurate materials which they object to. Over the years, the Wikipedia entries on "Wilhelm Reich", "orgone energy", "orgonomy", "orgone accumulator", "cloudbusting", and so forth, have occasionally shown some allegiance to facts, but typically degenerate after being invaded by Reich-hating vandals. Important factual materials then get erased, and substantial falsehood and even slander is inserted – usually by suspected professional "skeptic clubbers" and other "activists" with a personal grudge against Reich's ideas and discoveries. None of the revisionist scribblers use their real names when doing this, of course, and so the whole thing boils down to some foul "consensus", enforced by a slanderous cadre of angry anons who declare what is "truth" or "fact", based upon personal or political agendas. Those with real knowledge, with real jobs limiting how much time they can devote to daily monitoring of Wiki pages, are thereby silenced and censored out of the discussion.

Around the year 2000, I spent many hours correcting false statements and adding numerous documentary citations onto the Wilhelm Reich Wikipedia page,[33] detailing new scientific papers, academic dissertations, published articles in peer-reviewed journals and new books which supported Reich's biophysics. All the studies corroborating Reich, as found in the aforementioned *Nature* rebuttal article,[32] for example, were also added to the Wikipedia webpage on Reich.

Within a few months, all my entries, as well as additional factual materials provided by other authentic scholars with a good knowledge of Reich, were erased and deleted by some malcontent, replaced by words declaring that no such independent scientific verifications of Reich existed. Instead, additional libelous fabricated material was added in, so the falsehood of "Reich the crackpot" could be claimed without opposition. Several times, I and others made new corrections to the Wikipedia entries, only to have them repeatedly vanish shortly thereafter. The facts about Reich's life and work, his findings and those of his associates, his larger orgonomic theory, all the newer scientific evidence validating his findings and indicating their importance for psychology, biology, physics, cosmology and atmospheric science, etc., all of it was *fully and repeatedly deleted from the Reich-related entries on Wikipedia*.

By contrast, nearly every published item carrying foul lies and bad words about Reich continues to be given full voice on those same pages. Most recently, Wikipedia has begun including the newer lies and slanders from Turner's "Orgasmatron" book, repeating them in great detail. Photos of the serious and mature Wilhelm Reich, from his years as a research scientist and biophysicist, were at one time removed and substituted with a photo of a baby-faced individual who didn't even look like him, also aiming to denigrate the man and his research findings.

This abuse is cleverly constructed disinformation, specifically designed for half-educated, lazy and not-too-sharp readers. Some is practiced deceit, where fully false or irrelevant "citations" are given to support a typical slander that tears Reich down in serious ways. The given citation may point to a serious work, such as Myron Sharaf's *Fury on Earth: A Biography of Wilhelm Reich.*[34] On the surface, that seems "accurate" and "scholarly", the original statement vilifying Reich is "cited" and "referenced". It all seems fully believable if you don't bother to actually check the primary source reference. But if you do take the time to look it up, you find *absolutely nothing to support the malicious slander found on the Wikipedia page!*

This is a new "low" for the Reich-haters, to deliberately concoct phony citations to support a falsified biography of Reich. It is the worst of unethical conduct imaginable, similar to fabricating data in an experiment. But they get away with it, firstly because it is all done anonymously, and secondly because *they have the power to do so.* Truthful corrections simply get deleted, again and again, leaving only the lies on public display. Maybe tomorrow some lone scholar will spend hours to correct the Wiki pages on Reich and Orgonomy, but history suggests that it won't be tolerated for very long.

Reich's Orgonomy is not alone in being a target of such deliberate disinformation and abuse by organized advocacy groups dominating the pages of Wikipedia, making it an unreliable source for anything of controversy. Do an internet search on the words "Wikipedia Lies" or "Wikipedia Slander" to locate a plethora of independent websites put up by scholars and others on the subject, containing their own protests against the ant-army of scribblers who apparently have plenty of free time on their hands to spread lies and defamation, compared to those who actually work for a living.

Summary

From the above – the UK Channel 4 / Wag-TV slanders, the *Nature* magazine slanders, the *Wikipedia* slanders – we see the more prominent examples of what Reich called an *emotional chain reaction.* The nasty lies from the Turner "orgasmatron" book, which itself was never subjected to serious peer review or scholarly fact-checking, are merrily repeated. Such defamation also creeps into scientific circles, where it makes its way into print or onto internet websites, also without attempts at verification or fact-checking. This very much repeats the pattern seen in prior decades where the Mildred Brady slanders against Reich as published in popular magazines were later reproduced verbatim by hostile editors in the *Bulletin of the Menninger Clinic* and the *Journal of the AMA.* The FDA "investigation" was thereafter triggered, and one must ask if that might be the goal in the current wave of multiple public attacks – to trigger another government investigation or other "actions taken" against those "troublemakers" whose work increasingly verifies the accuracy and legitimacy of Reich's research findings on human sexuality and life-energy, as well as on Red Fascism.

Channel 4/Wag-TV were forced to comply with a British broadcasting inquiry, via the Ofcom organization, along with an implied threat of possible lawsuit under the British libel laws, before they would back down. After having done considerable damage to Reich by referencing the smutty writings and lies in the Turner "Orgasmatron" book, they retreated and indicated they would pull the offending segments from their *Sex-Researchers* program.

The editors of *Nature* magazine retracted at least some of their slanders of Reich, out of public embarrassment if not allegiance to truth or fact. They still apparently think the Turner book is something of authentic scholarship.

Wikipedia is impossible. There is no serious interest in facts, and a small army of angry "skeptic clubbers", unethical academics and organized advocacy groups, hiding their identities, scribble away to keep the lies going.

It should also be stressed, I have not detailed here the numerous "Book Review" articles repeating and glorifying the Turner slanders as they appeared in mainstream major newspapers in both print and online versions, since the "Orgasmatron" book was published in 2011. Those are listed in the online Emotional Plague bibliography,[35] and so will not be repeated here. One may also consider how each of the "reviews" triggered countless copycat articles, or repostings to other websites, dramatically expanding the outreach of the slanders. The collective repercussions of these attacks continue to reverberate.

6. Dusan Makavejev's Porno Film: *WR Mysteries of the Organism* *

"That one may smile, and smile, and be a villain."
William Shakespeare, *Hamlet*

Years before Mr. Turner wrote his *Adventures in the Orgasmatron*, the record for the worst and most damaging slander of Wilhelm Reich in the period after his death was found in the avant-garde porn film *WR Mysteries of the Organism*.[1] Produced in 1971 by Yugoslavian director Dusan Makavejev, *WR Mysteries* (hereafter *WRMO*) was shown in offbeat and university cinemas throughout the latter decades of the 20th Century. It was released as a VHS videotape in 2001, with a new DVD release and publicity campaign starting around 2006.

The film's first segments interspersed archive footage from Wilhelm Reich's early research, mixed with scenes of tender love-making. It began with a young man and woman off in the distance, removing their clothing and tenderly embracing in a love scene out in an open meadow under the blue skies, with wonderful emotive music. It was a naturalistic non-pornographic and deeply moving scene. A reasonably accurate narration from Reich's work accompanied the sound track in those initial scenes. Then director Makavejev suddenly inserted the out-of-context phrase *"fuck freely, comrades"*, followed by a shift to sexually degrading words and pornographic scenes with hysterical music and references to Marx and communism. The effect on the viewer was like a punch in the belly.

The film progressed from there through degenerated cartoonish or pornographic scenes which included:
- Adolescent narcissistic naked sexual displays.
- A former member of the *Fugs* 1960s rock band mock-masturbating the barrel of a machine-gun with a fake-blissful look on his face.
- The editor of *Screw* magazine being masturbated to erection by

---

* Originally internet posted in 2007.
     http://www.orgonelab.org/makavejev.htm

201

two young women of the "plaster-casters" porn-cult, after which they make a rubber dildo from the casting of his penis.
- A "scholarly" lecture from a "masturbation therapist".

The film jumped from one distortion of Reich to another. Mostly it was childish pregenital adolescent material, designed to titillate sex-frustrated people and elicit cheap anxious laughs. A sickening subplot continued on and off through the film of a female Communist Party member chasing after a sexually-dysfunctional ice-skating champion, who out of frustration towards the end of the film, chopped off her head with his ice-skates. The disembodied head then spoke to the film audience from a tabletop. Interspersed among these scenes were a few authentic and serious interviews with Reich's family, supporters and former associates, who had no idea how their contributions would be exploited.

Whatever serious issues that might have been suggested at the beginning of the film, or by the few rational interviews made with students or friends of Reich, were torn to shreds very quickly, leaving every decent person in the audience feeling nauseous. By this method, the name and work of Wilhelm Reich was falsely associated with sleazy pornography, sex-pathology and murder.

An earlier film review article of *WRMO* in the *Journal of Orgonomy*,[2] by John Bell and Barbara Koopman, revealed how the director Makavejev approached several of Reich's former associates and family members, eager to obtain documentary materials from them to use in his hit-piece, and to interview them. The article stated that Makavejev

> *"...asked them to cooperate with him on a documentary film he was making on the life and work of Reich. He presented himself as a Yugoslav democrat who had a deep interest in Reich's work and wanted to make a film that would set forth Reich's career and discoveries accurately. He was therefore accorded every courtesy and cooperation, and in turn gave many assurances that, despite his background, he was not a Communist; that he had read Reich's works extensively and understood the difference between freedom and license, love and fucking, and the primary and secondary drives; and that he would protect Reich and his work from the political and pornographic distortions to which they had so often been subject in the past.*

And finally we believed him -- but should we not have known better?" [2]

The Bell-Koopman report indicates how Makavejev used deception and trickery to obtain documentary materials and interviews from Reich's associates, claiming to be making an honest documentary for German television. None of it was true. The documentary footage and interviews acquired by deceit turned out to be the only valid part of the entire *WRMO* film. Without that, it would have been just another unnotable pornographic movie, mixed up with Marxist rantings.

In a February 2004 review of *WRMO*, I wrote the following:

"A Total Distortion of Reich's Work, James DeMeo, PhD, 2 February 2004

This film [WR Mysteries] was undertaken by Dusan Makavejev, who according to a 1971 film review article in the <u>Journal of Orgonomy</u>,[2] obtained original footage of Reich from the Reich Museum, and interviews with various individuals who knew Reich and who had followed up on his research, by posturing as a "friend of Reich". He then proceeded to mix that original footage with pornographic images designed to plunge a knife into the heart of everything Reich argued about, and stood for. Only a few of Reich's genuine friends had the forethought to grant permissions based upon their approval of the final film - and they promptly refused such permission. It caused an uproar among those who knew and understood Reich, but Makavejev got his footage and danced away a laughing man. This film has done more damage to Reich's name and legacy than any single item one might point to, by distorting Reich's excellent and important biophysical work on the <u>Function of the Orgasm</u> (see book of this title) into a malignant advocacy of "free sex for all". Reich would puke forever if he saw how his life's work was so badly twisted. "Fuck Freely" announces one of the heroes of this film, in a plot about an ice-skater who seduces, but then murders and decapitates one of the "revolutionary proletariat" female characters. A weak "plot" indeed, interspliced with porno images from the plaster-casters, the first male erection to appear on a US film,

In Defense of Wilhelm Reich

masturbation images, and people group-fornicating to cartoon music -- Makavejev's apparent idea of the "sexual revolution" -- all under a smiling wall portrait of Reich. "Freedom Peddling" is what comes to mind, and I can say with confidence that there is nothing of accuracy or authenticity about Reich in this film, aside from the interviews and short seconds of film from Reich's original archives, which were obtained by fraudulent means. Reich's work informs us that the pornographic character is a sexually-frustrated character, no less than the sex-negative moralist of the organized church (or mosque, temple, etc.), and that they are in fact mirror images of each other, flip sides of the same coin. "Brothels are built from bricks of religion" as the poet William Blake once said -- and Reich's clinical work put substance to this idea. And so when this film was shown at my university years ago, the theatre was flooded with all the sex-frustrated fraternity boys, come to hoot and ogle at the images of naked people. They learned nothing. Neither will anyone viewing this film, except perhaps how deceptive some Hollywood-types can be, even if they come from Yugoslavia."

More recently, another of Makavejev's pornographic efforts with the mis-title of *Sweet Movie* [3] ("Wretchedly Sick Movie" would be more accurate) was jointly released on DVD along with *WRMO*. Numerous positive reviews then appeared in film magazines and Hollywood internet sites, almost always very happy with both films, and misrepresenting Reich as the originator of their pornographic mentality and content.[4,5]

Under the headline *"Nocturnal Admissions"* one film review[5] stated that *Sweet Movie* created an uproar in Poland, where the very naked "actress" Anna Prucnal was banned from reentering the country. Another film review, given below, falsely claimed Reich inspired Makavejev's pornographic mentality.[6] This description clarifies why ordinary decent people would have rational objections to Makavejev's films, and why it is so inaccurate and upsetting to see Reich's name and work associated with them:

*"Like his WR: Mysteries of the Organism, Dusan Makavejev's controversial 1974 feature **Sweet Movie is firmly rooted in the principles of psychoanalyst Wilhelm Reich**. ... One*

begins with a trip to the "Miss World Virginity Contest," whose winner, Miss Monde 1984 (Carole Laure) is auctioned off to Mr. Kapital (Animal House's John Vernon) ...he sterilizes the terrified girl's body with rubbing alcohol and showers her in urine with his massive gold-plated penis, while an audience watches bemusedly through his bedroom window. She later escapes from her bridegroom, in a suitcase, and winds up at a wild Viennese commune whose participants indulge in public defecation and a food orgy that wraps with a massive display of gurgling, yakking, and vomiting. At the tale's conclusion, Miss Monde shoots a television commercial that involves writhing and masturbating in a giant vat of chocolate .... The second story involves a woman, Anna Planeta (Anna Prucnal) piloting a candy-filled boat down a river, with a massive papier-mache head of Lenin on the prow and a lover in-tow who is a refugee from the Battleship Potemkin. She eventually does a seductive striptease and seduces a pack of children, then makes love to her paramour in a vat of sugar and stabs him through the heart. Throughout the film, Macavejev includes shock cuts to Nazi autopsy footage and medical experimentation footage, some of which involves physical abuse of infants under the guise of "baby gymnastics." [emphasis added, JD][6]

Here's from another review:

"Sweet Movie -- a blitz of outrageous and nearly criminal offenses, cobbled onto a handful of silly dream-plots...Otto Muehl's regression therapy (in which members of Muehl's commune vomit and pee all over each other)...dinner plates of fresh shit, a castration in a vat of sugar..." [4]

For those who are unaware of the Muehl commune, it was a Viennese sado-pedo sex-cult called the *Aktions-Analytische Organisation (AAO)*, a spin-off of the 1960s European student's movement which fully distorted Reich's ideas on sexual freedom, turning them upside down and inside-out. By one internet account, *"Otto Muehl was in the sixtees a representive of 'Vienna-actionism' shocking the public. ... with the 'event of art' of public slaughter of gooses and pigs."* Muehl set up a commune in Austria, where he was arrested in 1990 for sexual abuse of the children in his commune,

In Defense of Wilhelm Reich

and sentenced to 7 years in prison. After getting out of prison, he set up a new commune in Portugal.

Of course, neither of Makavejev's twisted films, nor the bizarre and porno-sadistic antics of the AAO cult, were anything "rooted in the principles of psychoanalyst Wilhelm Reich". *They were the complete antithesis of Reich's clinical findings on what constituted healthy heterosexual genitality*, emphasizing love and tenderness, the protection of children from adult seducers and pedophiles, and gentle treatment of babies and children. The "connection to Reich" was fabricated by director Makavejev, whose film career was essentially built upon the *WRMO* film, the related semi-autobiographical book of similar title,[8] and a handful of lectures posturing as a self-proclaimed "expert on Reich."

I had the opportunity to hear one of Makavejev's lectures after a showing of his film, when I was studying at the University of Kansas back in the early 1980s. That was where the fraternity boys came to "hoot and holler" at the naked images in the film, as previously mentioned. In the after-film lecture, Makavejev had nothing of coherence or fact to say about Reich, but did proclaim, with some emphasis, *"Yugoslavia is the freest nation on the Earth!"* It did not seem tongue-in-cheek at all, but suggested a psychological disconnect, or pure Red propaganda, given how at that date Yugoslavia was still controlled by ruthless Communist Party pistoleros.

The damage done to historical fact and to the work of Reich can be illustrated in the following example. In the Rutgers University book *The Story of V: A Natural History of Female Sexuality,*[9] author and professor Catherine Blackledge referenced Reich's work through what she saw in the Makavejev *WRMO* film. She wrote:

> *"Reich's views of the importance of sexual pleasure were not shared by everyone, perhaps because of **his controversial exhortations to fuck freely. A propaganda film he made in his youth, Mysteries of the Organism, promoting what he called orgasmatherapy...**"*[9] (Emphasis added, JD)

In great error (or by deliberation?) Blackledge mis-attributed *Makavejev's* pornographic and mocking endorsement to "fuck freely", *to Reich!* She then repeated a full paragraph of Makavejev's ugly words from the film narration *as if they were a quote from Reich.*

There was an additional unethical side to this episode. When author Blackledge and her Rutgers University publisher were informed of these errors by letters from serious scholars, they basically shrugged off responsibility, *refusing to make any changes in their book.*[10] Thereafter, I posted the following Review at the book's Amazon.com website:

The Story of V: A Natural History of Female Sexuality (Hardcover) by Catherine Blackledge. REVIEW: Outrageous Lies In this Book, November 27, 2007.

The author gathers a lot of interesting material, but what stunned me was her apparent severe hatred for the important and pioneering sexual research findings of Dr. Wilhelm Reich, who in fact was one of the very first of the Freudians to speak favorably about female orgasm. But he also spoke about both male and female sexual impotence, identifying a lot of sexual behavior as rooted in sex-frustration and incapacity to achieve orgasm. This idea is hated by the S&M, "anything goes" advocates of "multiple sexualities". They hate Reich's guts. But instead of the author honestly making a criticism of Reich, citing from his publications that which she disagrees with, she instead quotes from the narrative of a pornographic film by the Yugoslavian director Dusan Makavejev, WR Mysteries of the Organism -- uniformly rejected and criticized by every authentic clinician and historian as a severe distortion of Reich -- claiming this was some kind of "propaganda film" MADE BY Reich. There is not a shred of fact in that claim. So are we to assume author Blackledge does her "research" by watching porno flics? And then mis-attributes what she sees to Reich??!! Merely because the director of the porno flic says so? Sloppy scholarship doesn't get any worse than that and it may have been a deliberate slander for all we know. What other falsehoods are in this book? It never should have gotten past the academic reviewers, but hey, the "new sexualities" allows anybody to claim anything, without worry of getting tagged by equally uncritical academic reviewers. This book is Not Recommended if you want a scientific discussion on sexuality. See Reich's "Function of the Orgasm" instead. It is still a light-year ahead of this nonsense.[11]

In Defense of Wilhelm Reich

Undoing the damage done to Reich and orgonomy by this one film, *WRMO*, will take decades. And the situation is even worse now with the arrival of the Turner "Orgasmatron" book, which reinforces all the same old falsehoods and slanders. People who stumble over those works are not likely to reevaluate what they have seen or read, especially since so many "authorities" appearing in the mainstream media and film industry continually reinforce the poisonous lie that the film is "rooted in the principles of psychoanalyst Wilhelm Reich".

Viewers of Makavejev films should therefore be aware of their deceptive background and content, that the pornographic and sick sexual themes of those films have nothing whatsoever to do with Reich, who was as much a critic of indiscriminate "free fucking" and an "anything goes" narcissistic poly-sexuality, as he was of rigid theocratic antisexual moralism.

One issue which still remains unresolved is the motivation of director Makavejev to make the original *WRMO* movie. The film was produced at a time, in the late 1960s and early 1970s, when Yugoslavia was under tight control of the Yugoslavian dictator Josip Tito and his Communist Party. In Makavejev's appeals to the American followers of Reich, to get documentary and interview materials necessary to help make the film, Makavejev told one lie after another about his plans for the film and his own interests in Reich. As noted above, he also claimed to be a "Yugoslav democrat" and "not a Communist".

These claims by Makavejev can be legitimately doubted, however, and it is possible the film was the result of a continuing deliberate Red Fascist conspiracy to do serious harm to Reich's research legacy, given Reich's strong criticisms of the Communist Party starting in the mid-1930s. How better to discredit Reich in the eyes of ordinary decent people than to distort him into an advocate of both communism and extreme pornographic filth? For example, Makavejev's film credits[13] include what appear to be several Communist Party propaganda items, such as the 1959 work *Sto je radnicki savjet? (What Is a Workers' Council?).*[12]

However, it is also possible he was "merely" going dirty-minded, and as seen in other instances of media distortion, he just said whatever nice words or lies were necessary to motivate people to cooperate with him, and to get the movie shots he needed to complete the film. And the truth be damned.

WRMO and *Sweet Movie* were banned in Yugoslavia during Tito's regime, and with added pressure from the Soviet Union he was exiled from reentering the nation until 1988.[13] The Communist Party bosses did not eject Makavejev because he had denigrated Wilhelm Reich, however. They were upset because the scenes of intensive sexually-twisted pornography were mixed up with their beloved Marx and Communism.[13,14] Tito is reported to have walked out of a screening of *WRMO*, calling it perverted.[14]

As mentioned previously, in 1981 when Makavejev lectured after showing the *WRMO* film in Lawrence, Kansas, I was there and heard him speak. He made condemning statements about the United States, and declared that the economically retarded Communist prison-state of Yugoslavia was some kind of paradise. Perhaps such public statements were designed to get him back in good graces with the new Yugoslav-Communist leadership, to lift the ban on his forced exile. Tito had died in May of 1980, but Yugoslavia remained in the iron grip of new Communist Party leaders for many years thereafter, a situation which gradually deteriorated that "freest nation" into fractious ethnic disputes, open war and genocides.

WRMO could have been thought-out Red-CP propaganda, to further destroy someone who knew and understood the Communist plague better than others, and hence whose writings posed a greater threat to their agendas, even though he was long dead. Or, it could simply be an expression of the director Makavejev having grown up in the harsh and sex-repressive communist Yugoslavia, but admiring the 1960s rebellion of youth culture as observed in the West from a distance. Both could be true, though the latter possibility does indicate a motivation which is rather obvious and in plain view. Surely, Makavejev must have felt and observed the powerful clash between the rigid communist system and people's authentic aspirations for freedom and sexual happiness. However, into this rational struggle of people breaking out from under the yoke of the Communist Party comes another possible clash, personified by Makavejev himself. The director's core biological push towards sexual love, and creative skills in the use of light and color, clashed with his own impotent sex-frustration and secondary layer attraction to dirty-minded scenes, very much like a comedian who always goes for cheap laughs by telling jokes about farting, getting drunk or stoned, or about women's breasts. It is a posture one particularly observes among artists who deliberately waste what talent they might have.

Any film amateur who can hold a camera straight is able to shock an audience with provocative scenes of nudity mixed with filth, to shove a camera lens into a dinner plate of feces. The mark of true genius is when the film as a whole touches upon something essential and real, exposing the deeper human emotions which make it important, and tell a story which has timeless roots. Makavejev never arrived at such a station, not by a long shot. He gained his brief notoriety only by inclusion of such towering figures as Wilhelm Reich in his ugly public display.

7. Orgone Flapdoodle: Enthusiast-Mystic Internet Nonsense

"Nothing in all the world is more dangerous than sincere ignorance and conscientious stupidity."
Martin Luther King, Jr.

Every scientific enterprise which digs into serious matters where an element of mystery persists, invariably runs into the problem where confused mystics or clever distorters steal away its central terms and findings, twisting them in strange or ridiculous ways, frequently to sell trinkets to the gullible. Archaeologists studying the ancient Egyptians, Mayans or Incas are driven nearly crazy by the numerous unserious references to "Atlantis" or "crystal skulls" they must field when dealing with the general public. Astronomy and space science suffer under the burden of those claiming with *absolute certainty* that there *has to be* an alien face on Mars, or that the Moon landings never happened, no matter what the evidence.

The modern fields of dark-matter and biophoton research, and quantum physics are hounded by various mysticisms and gadget-sellers misappropriating their terms, to the point that the physicist Murray Gell-Mann was driven to include a chapter on "Quantum Flapdoodle" in one of his books.§ Therein he complained about lay enthusiasts and misguided others, *"stringing together a series of terms and phrases from quantum physics and asserting that they explain something in our daily experience."* Here, I shall unapologetically echo Gell-Mann's complaint, and his term, to describe a similar problem in how Wilhelm Reich's *orgone energy* – an empirically demonstrated phenomenon with similarities to the *cosmic ether of space* or *dark matter* – has been misappropriated from its

---

§ Gell-Mann, Murray: *The Quark and the Jaguar: Adventures in the Simple and the Complex,* Chapter 12: Quantum Mechanics and Flapdoodle. Owl Books., 2002.

In Defense of Wilhelm Reich

roots in natural-scientific discovery and turned into a soapbox lingo for internet trinket hawkers. However, with respect to Reich, the damage is quite serious, as we are already struggling against 80 years of malevolent distortions from "skeptics", who consistently lie about the experimental record, trying to make the orgone energy sound ridiculous.

The flapdoodlers and ebay hawkers give the "skeptics" easy ammunition by which to dismiss Reich on the "ridiculous" side of the equation, as they have no interest to separate out the facts of what Reich actually discovered, versus what the orgone flapdoodlers believe and claim. In fact, when speaking about Reich, his detractors are quite happy to deliberately mix them up. Surely both the "skeptics" and the flapdoodlers enjoy freedom of speech, but so do I, and here I will exercise it, knowing full well that only a tiny fraction of the flapdoodle "believers" will be persuaded to self-critically undertake a rethink, and investigate the issues more seriously.

While the flapdoodlers use terms which sound like they might come from Reich, such as *"orgonite", "orgonium", "orgone generators", "chembusters"* and the like, none of those terms nor anything they make or sell, nor the theories they promote, have any foundation in the writings or discoveries of Reich. Nor can they be referenced to any of Reich's modern natural scientific investigators.

"Orgone Generators", for example, are plug-in devices or stand-alone electronic gadgets about the size of a radio, resembling negative-ion generators or old-style radionics instruments, except they are used to accomplish some rather farfetched goals. "Orgonite" and "Orgonium" are nearly identical compositions of transparent resins formed into different shapes – much like the 1970s "sand candle" fad – usually as fist-sized pyramids or cones or lumps, and into which are imbedded colorful metal shavings, BB pellets, crystals and other things. Some of them are quite eye-catching and might be classified as colorful works of art, but one could search forever in Reich's writings and never find even one single mention of them, nor anything similar.

I have several of these creations in my lab, purchased years ago for evaluations, to see if they really might constitute something new and important. However, nothing could be confirmed as to their claimed effects, other than a slight subjective radiation which is typical of nearly all kinds of matter, including from a metal cup, or the table on which it might be sitting. Our orgonite lumps are now used as paperweights. The claims for what they are supposed to do, however, go way beyond mere *objets d'art* or paperweights.

The orgonite/orgonium trinkets are advertised to attract all kinds of pleasant things into your life. You can buy "confidence building" lumps of orgonite, or "money attracting" orgonium pyramids, etc. By having that orgonite lump in your pocket, you are sure to attract a serious love partner, we are informed.

Also available are the "holy hand grenades" (or "HHGs", one of the original cone-shaped thingys) and "chembusters" – a large pail of resins into which metal shavings, BB pellets, crystals and multiple copper pipes are embedded.

The HHGs and chembusters have more practical uses, however, such as repulsing evil extra-dimensional demons, or demonic reptilian space invaders who are taking over the planet (seriously). Within the orgonite-chembuster cults, the leaders actually speak about how aircraft pilots, ground crews, and government officials have been kidnapped and replaced by *shape-shifting space reptilians* who are working in a grand conspiracy to spray poisonous "chemtrails" from civilian and military jet aircraft. Presumably, at some point, Earthlings will die off and the space frogs will take over. To the orgonite-orgonium, chembuster-chemtrail enthusiast, ordinary jet contrails are in reality *extraordinary* toxic "chemtrails". But you can now buy "HHGs" and "chembusters" to solve such deadly problems and do your bit to save humanity!

On eBay and YouTube one finds an even greater array of such fantastic hardware on display. One can find "orgone crystal power accumulators" (bearing no resemblance whatsoever to Reich's original orgone accumulator), "orgone pendants" and "pendant-beamers" (again, no connections to Reich), "orgone manifestors", "orgone chalice transmuters", "tachyon orgone Reiki massage throw blankets", "therapeutic orgone wealth prosperity and money pillows", "orgone generators for love and sex" or for "weight loss" or "...career, business and prosperity", along with "radionics rife orgone tesla HHG zappers", and even "orgone-powered Ouija spirit boards", all sold with Reich's name and terms stuck to them like so much glitter sprinkled on glue. *"Money! Wealth! Power! Sex!"* some of them shamelessly proclaim.

Sometimes they are marketed to farmers, to boost garden crop yields – and on that matter, as shown in prior chapters, the standard orgone accumulator does have quite a profound influence.

In Defense of Wilhelm Reich

But the orgonite lumps we have seen and tested have no experimentally-confirmed effects, and one does not expect it from gadgets originally made to repulse invisible shape-shifters or demons.

Some of the orgonite-chembuster hawkers are aware of this problem of credibility lost by referencing extraterrestrial demons or space frogs. So, a few years ago they began a program to clean up their language, and now latch on to current issues in modern environmentalism. Are you being zapped by microwave cell towers? Buy some orgonite lumps! You can then "gift" those lumps all around the cell tower and, presto zingo, problem solved. Worried about Monsanto and GMOs in your food? Put some orgonium lumps into your refrigerator! Worried about the CIA and the "geo-engineers" destroying the planet? You can now blame them for the overhead "chemtrails", buy a "chembuster" and fight back! And if you can't see the logic in all of this, well then you must be a CIA shape-shifter!

One group of "orgonite gifters" actually went to a drought region of Africa and began plunking their lumps into a large reservoir, trying to chase demons or bring rains, in a program they called "Orgonize Africa". The local cops found out about it, and not knowing what they were dumping into the precious water supply, arrested the "orgonizers", who landed in a dank prison cell for some months. The African authorities finally figured out what they were dealing with, and deported them. Of course such antics work to discredit legitimate and serious concerns about drought, microwave cell towers, GMOs and geo-engineering schemes. They also discredit Reich's name and work.

I would like to retain a sense of humor about this situation, but in fact I get e-mails from people asking me about this kind of thing every week, along with not a small amount of angry hate mail when they read a special webpage I put up which contains information similar to this chapter: (www.orgonelab.org/chemtrails.htm) Some of the most fanatical orgonite-orgonium-generator-chembuster advocates are also smoking pot and taking hallucinogens, which helps to understand the absence of self-critical capacities and frequently irrational and contactless behavior involved.

The orgonite-chemtrail advocates are frequently also heavily invested in other kinds of populist conspiracy theory, a common denominator which suggests a severe Oedipal hatred for home and country. This typically gets displaced as a primitive hatred of conservatives, Americans, Jews and capitalism – but of course never a dislike of totalitarian Islamic or Marxist murderers, nor of the atheistic pedophiles. One hopes that, at some point, they will put away the weed, tire of stirring epoxy, and move on in life to more serious matters.

Overall, and just like the defamatory slander from the "skeptics", the orgone flapdoodle distortions are broadcast loudly into the public sphere, and obscure the facts of Reich's discoveries from view. In some cases, honest people are simply diverted away from the more serious findings of Reich, as they do not know how his discovery anticipates that every kind of matter will carry a weak though variable charge of orgone energy, with a subtle radiant field. So they buy a lump of orgonite and feel a slight radiation – which is also possible to sense from a large metal cabinet or desk, or fluorescent light bulb if you are sensitive to energy fields. There are many materials and objects which possess active radiant energy fields, even magnets, which sensitive people can feel, or even see in darkened rooms. So in a manner, these lumps of materials are "on the track" of what Reich came upon in Oslo back in 1937. But they haven't made the leap into his findings from the mid-20th Century, much less to what is objectively and scientifically known today in the early 21st.

The stronger orgone charge and radiant field effects discovered by Reich are only possible by layering up ferromagnetic materials with dielectrics, as in the basic discovery of the orgone accumulator. In simplest form, this can be readily demonstrated with a plastic bag filled with steel wool pads, such as you can buy in any hardware store. Allow that bag of steel pads to rest on your open palms and the radiant charge will be generally felt by nearly everyone. The radiant field and charge of a properly-constructed Reich orgone blanket and accumulator is far greater.

In short, the internet "orgone gadget" sellers may claim to have advanced "far beyond Reich", but in fact their devices are retrograde and a giant leap backwards. Their gadgets are surely more colorful and sparkly than the "plain old-fashioned" orgone blanket or accumulator, but they simply do not function as authentic orgone accumulating devices.

A simple internet search on "orgone" will illuminate the problem, turning up far more weblinks promoting the flapdoodle than anything authentic or serious, and generally at the "top listings". A

In Defense of Wilhelm Reich

Google search of "orgone" in early 2013 turned up 1.6 million webpages, with perhaps only 1 out of 100 providing authentic information. I also made a similar survey of eBay, and found they were offering over 1600 items with "orgone" in the title – out of this number, no more than 1 or 2 percent used the word with accuracy, roughly paralleling the situation on the general internet.

As a help, I've created a special Resource Guide which provides the names and weblinks of the most authentic Reich/Orgonomy institutes and organizations, and the very few manufacturers of authentic orgone energy accumulators. Certainly there may be some we have missed who are not included, but it is a reasonably comprehensive and useful guide for the beginner to consult:
www.orgonelab.org/resources.htm

One could imagine, in the years around the time of the Wright brothers' famous flight at Kitty Hawk, there must have been a lot of people with broken legs, who had jumped off the barn roof with boards tied to their flapping arms, trying to fly like a bird. Possibly even the young Wrights may have considered or tried barn-jumping as a first step. But the difference between them and all the others was, that they actually studied birds, saw through the mystical flapdoodle, and kept working on the subject along empirical, experimental and natural scientific lines – even while the whole mainstream of science and journalism was happy to honk and bray at them in ridicule, like a herd of jackasses. Robert Goddard, inventor of the first liquid fuel rocket and whose discovery made space exploration possible, was also nearly destroyed by the honking and braying, which caused his financial support to dry up. The public was content to shoot off Chinese firecrackers and bottle-rockets, but they had no imagination and could only make a mockery of his work and vision, of controlled rockets going to the moon. He died nearly unknown, with the notable exception of certain German rocketeers, who expanded upon Goddard's designs and *were* taken seriously by the psychopath Adolf Hitler, out of which came the horrible V-weapons that rained death upon British civilians and allied troops during WW-II. Nobody told jokes about Goddard after those events, but he was dead by that time.

I could live more easily with the orgonite-orgonium pyramid makers than with the FDA, AMA or "skeptics". The flapdoodlers at least imagine and dream of what could be, albeit in a mystified and generally ineffectual manner. The FDA and "skeptic" types have no dreams or imagination at all, and can only work towards blocking or killing anyone who does! But one must nevertheless consider both expressions, of FDA/skeptics and mystic flapdoodlers, as a sign of our crazy and dangerous times.

In Defense of Wilhelm Reich

Appendix

CSICOP, Prometheus Books, Pornography and... the *Journal of Pedophilia*?! *

Prometheus Books, technically separate from CSICOP[§] but run by CSICOP Chairman Paul Kurtz and publishing many titles by CSICOP Fellows, sells books which range far beyond mere interests in "Scientific Rationalism", but which steer directly into pornography, and beyond. For example, the "Human Sexuality" section of Prometheus Books catalog is edited by CSICOP Fellow and *International Academy of Humanism* officer Dr. Vern Bullough, a prolific and controversial writer on sexual subjects who also wrote the introductions to books making uncritical, borderline-advocacy misportrayals of "consensual" adult-child sex and of human-animal sex. He also was listed as a member of the Editorial Board of the pseudo-scientific *Padika: The Journal of Pedophilia* on that organization's internet site.[1] Here is a sample of titles from the Fall-Winter 2000-2001 Prometheus Books Catalog:

- *The Horseman: Obsessions of a Zoophile,*
    by Mark Matthews, Introduction by Vern Bullough
- *Children's Sexual Encounters With Adults, A Scientific Study,* by C.K. Li, D.J. West and T.P. Woodhouse
- *The Q Letters: True Stories of Sadomasochism,*
    by Sir John

---

* Reprinted from *Pulse of the Planet* #5, 2002, p.67.

§ CSICOP = *Committee for the Scientific Investigation of Claims of the Paranormal*, recently rebranded as CSI, the *Committee for Skeptical Inquiry*. CSICOP/CSI is today's largest "skeptic" organization.

1. For more information, see the chapters by Edward Eichel in *Kinsey, Sex and Fraud* (Lochinvar-Huntington House Publications, 1990), and *Kinsey - Crimes and Consequences* by J.A. Reisman, et al. (Inst. for Media Education, 1998). Bullough wrote Introductions for *Dares to Speak : Historical and Contemporary Perspectives on Boy-Love*, J. Geraci, Ed. (Gay Men's Press, 1997), and for *The Horseman: Obsessions of a Zoophile*, by Mark Matthews (Prometheus Books, 1994).

219

In Defense of Wilhelm Reich

- *Whips & Kisses: Parting the Leather Curtain,*
  by Mistress Jacqueline
- *S&M Studies in Dominance and Submission,*
  by Thomas S. Weinberg
- *A Youth in Babylon: Confessions of a Trash-Film King,*
  by David Friedman & Don DeNevi
- *The X-Rated Videotape Guides: Volumes 1 - 8,*
  by Robert H. Rimmer
- *The X-Rated Videotape Star Index: Volumes 1 - 3,*
  by Patrick Riley
- *Raw Talent: The Adult Film Industry as Seen by its Most
  Popular Male Star,* by Jerry Butler
- *Dirty Talk: Diary of a Phone Sex Mistress,*
  by Gary Anthony & Rocky Bennett

One Prometheus title, PORN 101 (J. Elias, G. Brewer, V. Bullough, et al, Editors), was apparently drafted as a college textbook, highlighting a seminar of similar title (Porn 101: *Assimilating Pornographic Material in the Classroom*) which was presented at a recent "World Pornography Conference" sponsored by Bullough's *Center for Sex Research* at the University of California at Northridge. The "Conference" included as speakers many of the above authors, as well as various porn stars aping as "academics". Another Prometheus title, *Children's Sexual Encounters with Adults* gathers and unscientifically misrepresents the historical and cross-cultural evidence regarding childhood sexuality, to white-wash pedophilia under the guise of "scientific rationalism". Such books are promoted and sold on internet sites which openly endorse all kinds of sexual pathology, including pedophilia. All of the above titles were gathered on the last pages of the Prometheus catalog, as if to conceal their existence from the casual reader who might be interested in their other titles, but offended at the more outrageous pornography.

Given that Prometheus Books has several hundred titles on its list of publications – including many excellent classics and titles by well known authors including various CSICOP Fellows and officers, plus some very important books critical of authoritarian religious dogma – with so many other excellent manuscripts begging for a publisher, one can only wonder what the motivation is to include such extreme pornographic materials in their listing. The personal interests in these materials by some of America's leading "skeptics" appears quite clear. *It therefore does not appear accidental that CSICOP should also be one of the central-most sources of attack-and-ridicule disinformation directed towards Wilhelm Reich and his contemporary advocates.* It was Reich, after all, who wrote extensively about genuine sexual health and liberation, and about natural adolescent love within their own age groups, but who also roundly condemned the pornographer as being antisexual in nature, and pedophiles in particular as being deeply sexually sick, requiring police intervention to keep them away from children.

Reich observed: The pornographer destroys the more gentle and emotional-romantic side of sexuality just as surely as the church moralizer destroys the erotic-passion side of sexuality. Sexual health is composed of equal parts of both loving tenderness and eroticism. Both the church moralist and the pornographer express hatred towards natural loving sexuality in general, and towards the opposite sex in particular, differing only in the methods used to smash down natural heterosexuality. Unfortunately, the most vocal critics of the contemporary academic-pedophile movement are religious conservatives, who offer little more than their own brand of antisexual religious moralism as an alternative substitute (ie., the distortions that childhood sexuality does not exist, that contraception, abortion and divorce should be restricted again, and that adolescent lovers – ie., Romeo and Juliet – should be thrown into prison along with the pedophiles). Reich's sex-economic discoveries gain little support from either of these extremist camps, but rather stand as a deeper truth and common functioning principle underlying the two antithetical expressions, both of which are saturated with hatred towards love and sexuality.[2]

Postscript to the Postscript (2006)

Another author worthy to mention on this subject is Pierre Walter, who writes to destroy Reich in his role as a Reich "enthusiast". The problem is found in his books advocating pedophilia (under other names) as if Wilhelm Reich would also have approved of it – which is not true, as detailed in the prior chapters of this book (p.130-166) refuting the outrageous claims by author C. Turner. As

---

2. Also see Reich, Wilhelm: *Children of the Future: On the Prevention of Sexual Pathology*, Farrar, Straus & Giroux, NY 1983; DeMeo, James: "Wilhelm Reich's Discovery of Human Armoring" in *Saharasia*, 2nd Revised Edition, Natural Energy, Ashland, 2006, pp.17-46.

In Defense of Wilhelm Reich

previously noted, Turner's "orgasmatron" book slanderously depicted Reich as a pornographic pervert, even while in other forums, Turner vigorously defended the sexually pathological and criminal Kinsey agendas. Mr. Walter takes a different approach towards apparently similar goals. He once ran the "emosexuality.com" website, and otherwise promoted the concept of the "emosexual child" – which is to say, the child who is "seduceable" and adult-manipulated. He wrote various e-books on the subject: *The Minotaur Unveiled: The Constancy of Adult-Child Emosexual Attraction*,[3] *History of Childlove: Adult-Child Sexual Interaction*,[4] *The Emosexual Child*,[5] and *Emonics*.[6] In those works, Reich was presented as someone who might have agreed with Walter's poisonous ideas. More recently some of the same material appeared in a new print book *The Science of Orgonomy: A Study on Wilhelm Reich*.[7] Here, Walter provided meager and frequently wrong information on Reich's life-energy discoveries, but more objectionably continued to promote his own "emonics/emosexualism" under a fabricated and completely false "Reichian" banner.

3. Sirius-C Media Galaxy, Newark 2006. A new 2010 edition is subtitled: *Historical Assessment of Adult-Child Sexual Interaction*.
4. E-book, Sirius-C Media Galaxy, Newark 2005.
5. E-book, ipublica, Newark, 2005. see esp. pages 8-19.
6. *Emonics... Sexual Paraphilias*, Amazon CreateSpace, 2012.
7. Original edition: ipublica/Sirius-C Media Galaxy, Newark 2010, pp. 323-331. A newer edition is today available, which may be expunged.

Related Items to consult on the "Skeptics"

– "The Skeptics Are An Organized Hate Group" by Tim Bolen
  http://www.bolenreport.com/skeptics/Skeptics2/
      hate%20group.htm
– "The Skeptics: Who are they? What are they?" by Tim Bolen
  http://www.bolenreport.com/skeptics/index.htm
– "Skeptics Dragged into Court Over Rape, Mutilation and Death
    Threats" by Tim Bolen http://www.bolenreport.com/skeptics/
      Skeptics2Meryl%20Dorey%20death%20threats.htm
– "CSICOP and the Skeptics: An Overview" by George Hansen
  http://www.tricksterbook.com/
      ArticlesOnlineCSICOPoverview.htm
– Also see the "Additional Materials for Consultation" (p.245)

References and Notes

Introduction

1. Nitzschke, Bernd: "Psychoanalysis and National Socialism, Banned or Brought into Conformity? Break or Continuity?" *International Forum of Psychoanalysis* 12, 2003, p. 98-108.
2. Wolfe, Theodore P.: *Emotional Plague Versus Orgone Biophysics: The 1947 Campaign,* Orgone Institute Press, NY, 1948.
3. Blasband, Richard and Baker, Courtney (pseud. Rosenblum): "An Analysis of the United States Food and Drug Administration's Scientific Evidence Against Wilhelm Reich, Part I, the Biomedical Evidence", *Journal of Orgonomy,* 6(2):207-222, 1972; "...Part II, the Physical Concepts", *Journal of Orgonomy,* 6(2):222-231, 1972; "...Part III, Physical Evidence", *Journal of Orgonomy,* 7(2):234-245, 1973.
4. Martin, Jim: *Wilhelm Reich and the Cold War,* Flatland Books, Mendocino, CA, 2000. Revised edition, Natural Energy Works, Ashland, OR, 2013.
5. DeMeo, James: "New Information on the Persecution and Death of Wilhelm Reich", Chapter section in: *The Orgone Accumulator Handbook: Wilhelm Reich's Life-Energy Science and Healing Tools for the 21st Century, With Construction Plans,* Third Revised Edition, Natural Energy Works, Ashland, Oregon, 2010.
6. For a worst-case example, see DeMeo, James: "CSICOP, Prometheus Books, Pornography and the *Journal of Pedophilia?" Pulse of the Planet* #5, 2002, p.67, reprinted in the Appendix of this book.
7. See the online *Bibliography On Orgonomy*: http://www.orgonelab.org/bibliog.htm
8. Laing, R.D.: "Book Review: Function of the Orgasm", On *Wilhelm Reich and Orgonomy: Pulse of the Planet* 4:76-77 1993.

Chapter 2: Timeline

1. Sharaf, Myron: *Fury on Earth: A Biography of Wilhelm Reich*, St. Martin's-Marek, NY, 1983.
2. Reich, Wilhelm: "Über einen Fall von Durchbruch der Inzestschranke" (Breakthrough of the Incest Taboo), *Zeitschrift für Sexualwissenschaft* , VII, 1920.
3. Albrecht, Ulrich: *The Soviet Armaments Industry*, Harwood Academic Pub., 1993, p.13-17, 57, 62-67. Suvorov, Viktor: *The Chief Culprit: Stalin's Grand Design to Start World War II,* Naval Institute Press, Annapolis, 2008, p.17-18. Ericson, Edward E.: *Feeding the German Eagle: Soviet Economic Aid to Nazi Germany*, Praeger, 1999. Freund, Gerald: *Unholy Alliance: Russian German Relations from the Treaty of Brest-Litovsk to the Treaty of Berlin,* Harcourt Brace & Co., 1957. Gatzk, Hans W.: *Stresemann and the Rearmament of Germany*, Johns Hopkins Press, 1954.

Chapter References: 2. Timeline

4. Reich, Wilhelm: "Über Genitalität" (About Genitality), *Internationale Zeitschrift für Psychoanalyse*, IX, 1923.

5. Reich, Wilhelm: "Die therapeutische Bedeutung der Genitallibido" (The Therapeutic Importance of Genital Libido), *Internationale Zeitschrift für Psychoanalyse*, X, Vienna, 1924.

6. See the On-Line *Bibliography on Orgonomy* for the full listings. http://www.orgonelab.org/bibliog.htm

7. Reich, Wilhelm: "Die Rolle der Genitalität in der Neurosentherapie" (The Role of Genitality in the Treatment of Neurosis), *Zeitschrift für Ärztliche Psychotherapie*, IX, 1923.

8. Reich, Wilhelm: *Der Triebhafte Charakter: Eine psychoanalytische Studie zur Pathologie des Ich (The Impulsive Character: A Psychoanalytic Study on the Pathology of the Ego),* International Psychoanalytischer Verlag, #4 Leipzig/Vienna/Zürich 1925.

9. Reich, Wilhelm: "Weitere Bemerkungen über die therapeutische Bedeutung der Genitallibido" (More Comments on the Therapeutic Importance of Genital Libido), *Internationale Zeitschrift für Psychoanalyse*, XI, Leipzig/Vienna/Zürich 1925.

10. The details of Trotsky's terrorism and butchery, and his later deceit about it in various revisionist books and articles, only fully emerged in the latter part of the 20th Century. Much was concealed even by his enemy Stalin, as it would have exposed Lenin, Trotsky's boss, as complicit. Trotsky was basically Lenin's enforcer and executioner. For example, his violent fist-shaking "take no prisoners" speech to the angry mob in front of the Winter Palace in October 1917, defended only by a woman's brigade and troop of boy cadets, led to unnecessary rapine slaughter. This event was white-washed also in the later propaganda films of Sergei Eisenstein, which distorted the violent Bolshevik coup against freedom-seeking Russians into something heroic. Trotsky lied to the Kronstadt sailors defending the remnants of the early Russian democracy, promising them safety and fairness to get them to surrender, but then had most of them shot, unleashing the Red Army to pillage Petrograd. As Generalissimo, he sent the Red Army across large areas of Russia to terrorize and murder the remnants of the democratic Soviets, stomping out all opposition to Bolshevism in what was a prelude to the Red Terror and gulag death-camps. His conduct and advocacy of violent and global "War Communism" indicates, had he succeeded in the power struggle against Stalin, there would have been little difference in Soviet policies or behavior. For example see:
    - Volkogonov, Dimitri: *Trotsky: The Eternal Revolutionary*, Free Press, 2007
    - Sorokin, Pitirim: *Leaves from a Russian Diary*, Beacon Press 1950, (especially Trotsky's speech to the mob at the Winter Palace on p.61-62.)

11. Reich, Wilhelm: "Über die Quellen der neurotischen Angst (The Sources of Neurotic Anxiety)", *Internationale Zeitschrift für Psychoanalyse*, XII, #3, Leipzig/Vienna/Zürich 1926. Also in *International Journal for Psychoanalysis*, VII, p.381-391, London 1926.

12. Reich, Wilhelm: *Die Funktion des Orgasmus: Zur Psychopathologie und zur Soziologie des Geschlechtslebens (Function of the Orgasm)* Internationaler Psychoanalytischer Verlag, 1927.

Chapter References: 2. Timeline

13. Reich, Wilhelm: "Zur Technik der Deutung und der Widerstandsanalyse" (On the Technique and Interpretation of Resistance Analysis), *Internationale Zeitschrift für Psychoanalyse*, XIII, Leipzig/Vienna/Zürich 1927.
14. Reich, Wilhelm: People In Trouble, Orgone Institute Press, Maine, 1953.
15. Reich, Wilhelm: "A Criticism of Recent Theories of the Problem of Neurosis", *Internationale Zeitschrift für Psychoanalyse*, IX, p.227-240, 1928.
16. Reich, Wilhelm: "Onanie im Kindesalter (About Genital Self-Satisfaction in Children)", *Zeitschrift für Psychoanalytische Pädagogik*, No.4, 5 & 6, 1928.
17. Reich, Wilhelm: "Über Charakteranalyse (On the Technique of Character Analysis), Internationale Zeitschrift für Psychoanalyse, XIV, 1928.
18. Sharaf 1983, ibid p.132, summarizing Reich.
19. Reich, Wilhelm: "Die Stellung der Psychoanalyse in der Sowjetunion. Notizen einer Studienreise in Russland (The Position of Psychoanalysis in the Soviet Union: Notes from a Study Tour in Russia)", *Psychoanalytische Bewegung,* I(4):358-368, November/December 1929.
20. Reich, Wilhelm: *Dialektischer Materialismus und Psychoanalyse (Dialectical Materialism and Psychoanalysis)*, Sexpol Verlag, Denmark 1934.
21. Reich, Wilhelm: "Der genitale und der neurotische Charakter (The Genital Character and the Neurotic Character)", *Internationale Zeitschrift für Psychoanalyse*, XV, Vienna 1929.
22. Reich, Wilhelm: *Sexualerregung und Sexualbefriedigung (Sexual Excitation and Sexual Satisfaction),* Muensterverlag, Vienna 1929.
23. Reich, Wilhelm: "Sexualnot der Werktätigen und die Schwierigkeit sexueller Beratung (The Sexual Misery of the Working Masses and the Difficulties of Sexual Counselling)", *Sexualnot und Sexualreform: Verhandlung der Weltliga für Sexualreform,* abgehalten zu Wien am 16-23 Sept. 1930, Vienna 1931, p.72-87.
24. Reich, Wilhelm: "Über kindliche Phobie und Charakterbildung (Character Formation and the Phobias of Childhood)", *Internationale Zeitschrift für Psychoanalyse*, XVI(3/4):353-362, Vienna 1930.
25. Reich, Wilhelm: *Geschlechtsreife, Enthaltsamkeit, Ehemoral: Eine Kritik der bürgerlichen Sexualreform (Sexual Maturity, Abstinence and Marital Fidelity: A Criticism of Civil Sexual Reform)*, Muensterverlag, Vienna 1930.
26. Malinowski, Bronislaw: *The Sexual Life of Savages in North-Western Melanesia*, George Routledge & Sons, London 1929.
27. Reich, Wilhelm: "Über den epileptischen Anfall" (About Epileptic Seizures), Internationale Zeitschrift für Psychoanalyse, XVII, Vienna 1931, p.263-275.
28. Sharaf 1983, ibid p.162-163.
29. Kautsky, Karl: Article attacking Reich's booklet "Zur Kritik der Sexualreform (To the Critic of Sexual Reform)" in the left-wing *Freiheit* No.7, Berlin, 1931.
30. Reich, Wilhelm: "Der masochistische Charakter: Eine sexualökonomische Widerlegung des Todestriebes und des Wiederholungszwanges (The Masochistic Character: A Sex-Economic Refutation of the Death Instinct and Repetition Compulsion)", *Internationale Zeitschrift für Psychoanalyse*, XVIII, Vienna 1932.
31. Reich, Wilhelm: *Der Einbruch der Sexualmoral: Zur Geschichte der sexuellen Ökonomie: Zweite Ergänzte Auflage (The Invasion of Sexual Morality: A*

History of Sexual Economy), Second Supplemented Edition, Sex-Pol Verlag, Copenhagen, 1935.

32. Reich, Wilhelm: *Der Sexuelle Kampf der Jugend (The Sexual Struggle of Youth),* Verlag für Sexualpolitik, Berlin 1932. Partial English translation as revised by Reich in *Children of the Future,* Farrar Straus & Giroux, 1983.

33. Bernfeld, Siegfried: "Die kommunistische Diskussion um die Psychoanalyse und Reich's 'Widerlegung der Todestriebhypothese' (The Communist Discussion in Psychoanalysis and Reich's 'Refutation of the Death Instinct')", *Internationale Zeitschrift für Psychoanalyse,* XVIII, 1932.

34. Reich, Wilhelm: "Abschließende Bemerkung zur 'Gegenkritik' Bernfelds 'Die kommunistische Diskussion um die Psychoanalyse und Reichs 'Widerlegung des Todestriebhypothese'" *Internationale Zeitschrift für Psychoanalyse,* XVIII, Vienna 1932.

35. DeMeo, James: *Saharasia: The 4000 BCE Origins of Child-Abuse, Sex-Repression, Warfare and Social Violence, In the Deserts of the Old World,* Natural Energy Works, Ashland, Oregon 1998. (Revised and updated second edition in 2006).

36. anon.: Notice prohibiting distribution of Reich's books, *Roter Sport (Red Sport),* a publication of *Fichte,* the Berlin-Brandenburg sports organization of the Communist Party of Germany (KPD), 5 Dec. 1932.

37. Reich, Wilhelm: *Massenpsychologie des Faschismus (Mass Psychology of Fascism),* 1st Edition, Sexpol Verlag, 1933.

38. Reich, Wilhelm: *Charakteranalyse: Technik und Grundlagen für studierende und praktizierende Analytiker (Character Analysis: Principles and Technique for Psychoanalysts in Practice and in Training),* 1st Edition, Sexpol Verlag, 1933.

39. For example, see: Baker, Elsworth F.: *Man In The Trap,* Macmillan, NY, 1967; Von Uexkill, Theodore: Chapter on "Body Oriented Psychotherapy" with generous discussion of Reich, in *Psychosomatische Medizin,* Urban & Schwarzienberg, Munich, 1990; Crist, Peter and Schwartzman, Richard: "Medical Orgone Therapy", in Allison, Nancy, ed. *The Illustrated Encyclopedia of Body-Mind Disciplines*; Rosen Publishing Group, NY, 1999; Herskowitz, Morton: *Emotional Armoring: An Introduction to Psychiatric Orgone Therapy,* Transactions Press, NY 1998.

40. From a "Contributor to the German Association Conference", or *Unity Association Conference,* held on 29 Jan.1933, as quoted within a pirate edition of Reich's *The Sexual Struggle of Youth,* Socialist Reproductions, London, c.1972.

41. anon.: "Bolschewismus oder Deutschland?", *Völkischer Beobachter,* 2 March 1933.

42. Wolff, Wilfried: *Max Hodann (1894-1946). Sozialist und Sexualreformer,* Brockl Verlag, Germany 1993.

43. It is estimated that over 40 different German agencies drew up lists of offensive publications to be banned, numbering around 4,100 in total. See Leonidas Hill, "The Nazi Attack on Un-German Literature, 1933-1945," in *The Holocaust and the Book: Destruction and Preservation,* Edited by Jonathan Rose, Univ. of Massachusetts Press, Amherst 2001. While at

present we do not have 100% certainty, it is safe to assume Reich's books were burned at some or most of the many nationwide Nazi book burning events of 1933 and thereafter. It was his writings and speeches which led to his being denounced in the Nazi *Volkischer Beobachter* in March 1933, and his name appearing on a Nazi/Prussian Expulsion order in May of that same year. Later in 1935, the Gestapo ordered the banning of his books, as detailed in the text, which presumes their destruction when copies would be located, seized and impounded. Hill notes the chaotic situation at the time of the May 1933 Nazi euphoria: *"Although on 10 May 1933 Nazis chanted the 'Feuersprüche' [fire incantations], nine objectionable characteristics of books followed by the names of some of the authors, while throwing their books on bonfires, few knew the other books that were burned. Whereas anyone aware of the Nazis' views might guess quite accurately which volumes in his or her own library were suspect or condemned, the bookseller Hilde Wenzel knew that not all the books burned on 10 May 1933 were actually banned. In fact, until 1935 the Prussian police had not banned any of the volumes named at the burnings, not even Remarque's <u>All Quiet on the Western Front</u>. The regime did not have a master plan when they took office and did not soon design one. For some years Nazi bannings and seizures of books in the fifteen German states were uncoordinated and carried out by many agencies at different levels of government.*

The book burnings were a publicity stunt devised by one Nazi student organization... to curry favor with the government. ... no government ministry officially approved of the 'blacklist'... first sent to the students on 1 May and frequently extended thereafter." Hill, ibid, p.13.

44. Martin, James: *Wilhelm Reich and the Cold War*, Flatland Books, Mendocino, CA, 2000, p.95-106. Also see: Nasselstein, Peter: *Der Rote Faden: Genosse Thomas*, (Teil I + II), Internet Posted:
    http://nachrichtenbrief.wordpress.com/2011/07/16/der-rote-faden-genosse-thomas-teil-1/
    http://nachrichtenbrief.wordpress.com/2011/07/17/der-rote-faden-genosse-thomas-teil-2/
45. anon.: Article attacking Reich's *Mass Psychology of Fascism*, in *Arbejderbladet (The Daily Worker)*, Copenhagen 1933. In Siersted 1993, see citation 43, p.53.
46. Siersted, Ellen: "Wilhelm Reich in Denmark (1933-1939)," in *On Wilhelm Reich and Orgonomy: Pulse of the Planet #4*, James DeMeo, Editor, Ashland, Oregon 1993, p.44-69. Also see: Leistikow, Gunnar: "The Fascist Newspaper Campaign in Norway", *Int. J. Sex-Economy & Orgone Res.*, I:266-273, 1942.
47. Nitzschke, Bernd: "Psychoanalysis and National Socialism, Banned or Brought into Conformity? Break or Continuity?" *International Forum of Psychoanalysis* 12, 2003, p. 98-108. Nitzschke documents how the *German Psychoanalytic Society (Deutsche Psychoanalytiker Gesselschaft)* in 1933 began secretly expelling its Jewish members, and by 1936 formally merged with the *German Institute for Psychological Research and Psychotherapy (Deutsches Institut für psychologische Forschung und Psychotherapie)*, also called the *Göring Institute* given that its Director was Mathias Heinrich

Göring, a cousin of the Nazi Reichsmarshal Hermann Göring. M.H.Göring wrote *"the Society (for psychotherapy) assumes of all members who are active as writers or speakers that they have worked through in all scientific earnest Adolf Hitler's fundamental work 'Mein Kampf' and that they acknowledge it as the basis of their work."* Carl Jung was the Editor of this Society's *Zentralblatt für Psychotherapie*, in which for the very first issue he wrote: *"That there are actual differences between Germanic and Semitic psychology has long been known to intelligent people. These differences are no longer going to be obliterated."* Wilhelm Reich was one of the few IPA members to protest against these developments. (See: Translator's footnote, p.127 of Reich, Wilhelm: *Function of the Orgasm,* Orgone Institute Press, New York 1942.)

48. Reich, Wilhelm: "Abhandlungen zur personellen Sexualökonomie - Der Orgasmus als elektrophysiologische Entladung (Paper on Human Sex-Economy: The Orgasm as an Electrophysiological Discharge)", *Zeitschrift für Politische Psychologie und Sexualökonomie,* I:29-43, 1934.

49. Reich, Wilhelm: "Abhandlungen zur personellen Sexualökonomie - Der Urgegensatz des vegetative Lebens (Paper on Human Sex-Economy: The Basic Antithesis of Vegetative Life)", *Zeitschrift für Politische Psychologie und Sexualökonomie,* I:125-142, 1934.

50. Reich, Wilhelm: *Dialektischer Materialismus und Psychoanalyse (Dialectic Materialism and Psychoanalysis)* Sexpol Verlag, Denmark 1934.

51. Reich, Wilhelm (pseud. Parell, Ernst): "Was ist Klassenbewußtsein? (What is Class Consciousness?)", *Zeitschrift für Politische Psychologie und Sexualökonomie,* I:16-29; 90-107; 226-255, 1934.

52. Martin: *Wilhelm Reich and the Cold War,* ibid. p.288, 299-300.

53. See Document 20, *"Memorandum on Trotskyists and Other Hostile Elements in the Emigre Community of the German CP, Cadres Department"*, dated 2 Sept. 1936, in the Yale University Archives:
    http://www.yale.edu/annals/Chase/Documents/doc20chapt4.htm
    This document is also partly reproduced as "Document 17" in *Enemies within the Gates? The Comintern and the Stalinist Repression, 1934-1939,* by William J. Chase, Yale Univ. Press 2001, p.164-174. The reference to 3000 European emigres appearing on a detention (death) list, drafted by the various European CPs, is found in Chase, p.162. While Reich was never a follower of Trotsky, he did have a letter exchange and possible meeting with him. That alone, or just the mere accusation of it, would have been sufficient for Reich's name, and that of one of his Denmark-Norway contacts, Otto Knobel, to appear on the official NKVD list. Knobel's offense was having been a known associate of Reich, indicating Reich was the primary target and offender. The document carried notations on others who had already been apprehended and sent into prisons or off to the Siberian gulag, or executed. Knobel was in fact later arrested by the NKVD and imprisoned, or "disappeared" (executed).

54. Fenichel, Otto: *119 Rundbriefe (1934-1945),* Johannes Reichmayr, Elke Mühlleitner, Editors, Stroemfeld Verlag, Frankfurt am Main, 1998.

55. Reich, Wilhelm: "Überblick über das Forschungsgebiet der Sexualökonomie

Chapter References: 2. Timeline

(Overview of Sex-Economic Research)" *Zeitschrift für Politische Psychologie und Sexualökonomie, II:5-13, 1935.*

56. Reich, Wilhelm: "Der Ausschluss Wilhelm Reichs aus der Internationalen Psychoanalytischen Vereinigung (The Expulsion of Wilhelm Reich from the International Psychoanalytic Association)", *Zeitschrift für Politische Psychologie und Sexualökonomie*, II:54-61, 1935.

57. Item # 41230/35 11 2 B, later published in *Deutsches Reichsgesetzblatt*, 13 April and 7 May, 1935)

58. Reich, Wilhelm: "Psychischer Kontakt und vegetative Strömung (Psychic Contact and Vegetative Streaming)", *Zeitschrift für Politische Psychologie und Sexualökonomie*, 3, 1936, Sexpol Verlag, Kopenhagen.

59. Reich, Wilhelm: *Die Sexualität im Kulturkampf*, Sexpol Verlag, Kopenhagen 1936.

60. Reich, Wilhelm: (pseud. Ernst Parell) "Der Kulturpolitische Standpunkt der Sexpol", *Zeitschrift für Politische Psychologie undd Sexualökonomie* 3(1-2):3-4, 1936.

61. Reich, Wilhelm & Neill, A.S.: *Record of a Friendship: The Correspondence of Wilhelm Reich and A.S. Neill*, edited by Beverley Placzek, Farrar, Straus & Giroux, London 1982.

62. Reich, Wilhelm: "Experimentelle Ergebnisse über die elektrische Funktion von Sexualität und Angst (Experimental Investigation of the Electrical Function of Sexuality and Anxiety)" and "Orgasmusreflex, Muskelhaltung und Körperausdruck (Orgasm Reflex, Muscle Posture and Body Language)", *Klinische und Experimentelle Berichte* #4 and #5, Sexpol Verlag, Copenhagen, 1937.

63. Reich's critics and attackers were legion. For a full list of their published attacks, see: http://www.orgonelab.org/bibliogPLAGUE.htm

64. Lackner, Stephan: "Ein moderner Ketzer (A Modern Heretic)", *Das Neue Tagebuch (The New Diary)*, Paris, Vol.5, No.6, 1937, p.140-141. Reich wrote a letter to Lackner in response to this article, on 9 April 1937, suggesting his agreement with Lackner. He did not challenge Lackner's points that Marxism was incompatible with his sex-economic work, or that he would be (or was) moving away from Marx. See. Laska, Bernd: *Wilhelm Reich Blätter*, 2/80, p.67-85, and 2/81, p.65-70. Also see Nasselstein, Peter: "Der Rote Faden: Auf dem Weg in die Arbeitsdemokratie", http://nachrichtenbrief.wordpress. com/2012/09/23/der-rote-faden-auf-dem-weg-in-die-arbeitsdemokratie/

65. Reich, Wilhelm: *Die Bione: zur Entstehung des vegetativen Lebens, Experimentelle Berichte (Clinical and Experimental Report)* #6, Institut für Sexualökonomische Lebensforschung, Sexpol Verlag, Oslo/Kopenhagen/ Zürich, 1938.

66. duTeil, Roger: "Leben und Materie: Drei Versuchsreihen (Three Series of Experiments Based on the Tension-Charge Principle)", Communication to the Natural Philosophy Society, Nice, and Academie des Sciences, Paris, France, 1938. Contained in Reich *Die Bione* 1938, ibid.

67. Malinowski letter supporting Reich, published 12 March 1938 in *Dagbladet*, Norway. Reprinted in *Reich Speaks of Freud*, Mary Higgins & Chester Raphael, Editors, Farrar, Straus & Giroux, NY 1967, p.219.

68. A.S. Neill letter supporting Reich, in Sharaf 1983 ibid, p.232, written 25 June 1938, "A Warm Defense of Reich", *Dagbladet*, Norway.

69. DeMeo, James: "Contemporary Bion Research and Related Findings in Biology", presented to the *Second International Conference on Pleomorphism in Health and Disease*, 19-20 October 2000, Ashland, Oregon. http://youtu.be/-PVnS72IIY8

70. Reich, Wilhelm: "Bion Experiments On The Cancer Problem," *Klinische und Experimentelle Berichte (Clinical and Experimental Report)*, #7, p.3-25, Sexpol Verlag, Rotterdam, Rotterdam-Oslo-Copenhagen, 1939.

71. Reich, Wilhelm: "Drei Versuche am Statischen Elektroskop (Three Experiments with the Static Electroscope)," *Klinische und Experimentelle Berichte (Clinical and Experimental Report)*, #7, 26-27, Sexpol Verlag, Rotterdam, Rotterdam-Oslo-Copenhagen, 1939.

72. Neill, Alexander S.: *The Problem Teacher*, Herbert Jenkins, London 1938. Neill, Alexander S.: *The Problem Family*, Herbert Jenkins, London 1948.

73. Reich, Wilhelm: "Die natürliche Organisation der Arbeit in der Arbeitsdemokratie". *Politisch-Psychologische Schriftenreihe der Sex-Pol*, Nr. 4, Oslo: Sexpol-Verlag, (Jan.) 1939.

74. For example, see the following items by Wilhelm Reich:
- "Weitere Probleme der Arbeitsdemokratie". *Politisch-psychologische Schriften* No. 5, Europe (Rotterdam: Sexpol-Verlag), April 1941.
- In the 1943 issue of *Int.Journal of Sex-Economy and Orgone Research:*
"Give Responsibility to Vitally Necessary Work". "The Biological Miscalculation in the Human Struggle for Freedom", "Work Democracy Versus Politics", "The Natural Forces for the Mastery of the Emotional Pest".

75. Flexner, Abraham: *Medical Education in the United States and Canada: A Report to the Carnegie Foundation for the Advancement of Teaching*, Boston 1910.

76. Walker, Martin: *Dirty Medicine: Science, Big Business and the Assault on Natural Health Care*, Slingshot Publications, London 1993. Carter, James P.: *Racketeering In Medicine: The Suppression of Alternatives,* Hampton Roads, Oakland 1992. DeMeo, James: "Natural Healing *Living* Waters", Chapter section in: *The Orgone Accumulator Handbook: Wilhelm Reich's Life-Energy Science and Healing Tools for the 21st Century, With Construction Plans,* Third Revised Edition, Natural Energy Works, Ashland, Oregon, 2010.

77. Baldwin, Roger N.: "Freedom in the USA and the USSR", *Soviet Russia Today*, 3 Sept.1934, p.11.

78. *Discover the Networks. Roger Baldwin,* online resource of the David Horowitz Center for Democracy:
http://www.discoverthenetworks.org/individualProfile.asp?indid=1579

79. Donohue, William: *The Politics of the American Civil Liberties Union,* Transaction Publisher, 1985, p.45.

80. *Fourth Report of the Senate Fact-Finding Committee on Un-American Activities, 1948 : Communist Front Organizations,* California State Senate, Sacramento. Largely reprinted in: Dilling, Elizabeth: *The Red Network, A 'Who's Who' and Handbook of Radicalism for Patriots,* self-published 1934.

81. *Radio Highlights 1939-1957.* Internet. A search using the following weblink

Chapter References: 2. Timeline

on "Arthur Hays" brings up 39 item, most of which are relevant.
http://www.jjonz.us/RadioLogs/pg02_radioHighlights.htm

82. Hays, Arthur G.: *What I Saw In Russia* Haldeman-Julius Publications 1927.

83. Martin *Wilhelm Reich and the Cold War*, 2000, ibid., Chapter III and p.134. Wilmers, Mark-Kay: *The Eitingons: A Twentieth-Century Story,* Verso 2010, p.89-90.

84. Hays, Arthur G.: *City Lawyer: Autobiography of a Law Practice by Arthur Garfield Hays*, Simon & Schuster, NY 1942, p.339, 385-387. The book is dedicated to three fellows, including Roger Baldwin.

85. *Longines Chronoscope*: TV Broadcast 7 Dec.1951. Arthur Garfield Hays interviewed by Victor Riesel and William Bradford Huie.
http://www.youtube.com/watch?v=cOcgUzKytQU

86. "Consumer Group Formed" *New York Times*, 6 Feb. 1936. *Third Report, Senate Investigating Committee on Education Textbooks*, California Legislature, 1948 Regular Session, p.80. Also see An Inventory to the Records of Consumers' Research, Inc.: General Files: Series 16-27, January 1995, Special Collections and University Archives, Rutgers Univ. Libraries. Index online, do term search for "Brady".
http://www2.scc.rutgers.edu/ead/manuscripts/consumers_gen1b.html

87. *Senate Fact-Finding Committee*, 1948. See citation 80 ibid, p.167, 392. Kallet published in the *Daily Worker* under the pseudonym of "Edward Adams". Consumers' Union was removed by Congress from their list of subversive organizations only after 1953. See *Guide to Subversive Organizations and Publications,* Committee on Un-American Activities, 85th Congress, 1st Session, House Document 166, 1957, p.141.

88. Brady, Robert: *The Spirit and Structure of German Fascism*, with a foreword by Harold J. Laski. Left Book Club, 1937, p.324.

89. *State Department Loyalty Investigations, Hearings, Committee on Foreign Relations, US Senate,* 81st Congress, 2nd Session, March-June 1950, p.167-170.

90. Findings and Recommendations of the Personnel Security Board in the Matter of Dr. J. Robert Oppenheimer, US Atomic Energy Commission, 27 May 1954. http://avalon.law.yale.edu/20th_century/opp01.asp

91. Letter from Robert Oppenheimer to Mildred Edie Brady, 29 May 1938. in *Robert Oppenheimer: Letters and Recollections*, by R. Oppenheimer, Alice Smith, Charles Weiner, Stanford University Press, 1995, p.205.

92. "Dies Demands Henderson be Fired as Red", *Associated Press,* 8 Sept. 1941.

93. Russell, Edmund: *War and Nature*, Cambridge Univ. Press, 2001, p.89. Hapgood, Norman: *Professional Patriots*, Boni 1927.

94. Parrish, Michael: *Felix Frankfurter and His Times*, Free Press 1982, p.5,160. Frankfurter included a work by the British communist Harold Laski, *The American Democracy* (Viking Press, NY 1948), in a short list of his favorite books. (See Barrett, below.) In that book, Laski condemned American conservatives and capitalists as racists, Jew-haters and Nazis. He praised socialism, Marx and Lenin, denigrated anti-communism, and spoke favorably of Trotsky, the Soviet spy Henry Wallace, and communist-run magazines like *The Nation* and *New Republic*. (see *American Democracy* p.98, 119, 250-51,

Chapter References: 2. Timeline

254, 643-45, 649-52, 669, 743) No mention was made by Laski of the original and democratic *February* Russian revolution nor of Kerensky or the Russian Provisional Goverment, nor of Bolshevist or Lenin's or Trotsky's crimes, the Ukrainian genocide, etc. Laski also published *The Communist Manifesto: Socialist Landmark. An Appreciation for the Labour Party, Together With Original Texts and Prefaces* in that same year (Geo. Allen and Unwin, London 1948), in which he heaped praise upon Marxist totalitarian agendas. Laski, like Frankfurter, also published various pro-Soviet articles in *New Republic* and *The Nation.* He dedicated to Frankfurter one of his earlier books, *Authority in the Modern State* (Yale Univ. Press, 1919). In the context of the Laski-Frankfurter friendship and ideological alliance, it is therefore not surprising to learn that, duringin his days as Harvard Law Professor, Frankfurter would bitterly complain about the *Due Process* clauses of the 5th and 14th Amendments of the US Constitution, which trace back to the Magna Carta and English Common Law, providing all citizens with guarantees of *rights to life, liberty and property,* under the *due process of law.* *"The due process clauses ought to go"*, he stated in an unsigned 1 October 1924 *New Republic* article with the alarming title "The Red Terror of Judicial Reform". His basic argument was towards an unfettered Big Brother federal government where the Supreme Court should have arbitrary power over the States via the "commerce clause". This is exactly what happened with the 1938 Roosevelt empowerment of the FDA, giving it police powers over State and citizen activity under the commerce clause, in actions that nearly 20 years later led to the burning of Wilhelm Reich's books. During WW2, at the time of his Supreme Court appointment, Frankfurter also tried to persuade Roosevelt to share the secrets of the atom bomb with the Stalin. See:
- Frankfurter, Felix: "The Red Terror of Judicial Reform", in *Felix Frankfurter on the Supreme Court, Extrajudicial Essays*, Philip Kurland, Editor, Belknap/Harvard University Press, Cambridge 1970, p.158-167. (Also in Frankfurter, Felix: *Law and Politics: Occasional Papers of Felix Frankfurter*, 1913-1938, Capricorn Books 1962, p.10-16.
- Barrett, J.Q.: "Supreme Court Justices on their Book-Reading, 1948-1949," unpublished manuscript, p.8. (Available on-line)
- Ekirch, Arthur.A.: "Harold J. Laski: The Liberal Manqué or Lost Libertarian?" *J. Libertarian Studies*, VI(2):139-150, 1980.
- Persico, J.E.: *Roosevelt's Secret War*, Random House, NY, 2001, p.339-340.
.   - Also see citations 160 and 164 below.
95. Haynes, John Earl & Klehr, Harvey: *In Denial: Historians, Communism and Espionage,* Encounter Books, 2005. Haynes, John Earl & Klehr, Harvey: *Venona: Decoding Soviet Espionage in America,* Yale Univ. Press, 1980.
96. Reich, Wilhelm: *American Odyssey: Letters and Journals 1940-1947,* Farrar, Straus & Giroux, NY 1999, p.38.
97. DeMeo, James: *The Orgone Accumulator Handbook: Wilhelm Reich's Life-Energy Science and Healing Tools for the 21st Century, With Construction Plans,* Third Revised Edition, Natural Energy Works, Ashland, Oregon, 2010.
98. Reich, Wilhelm: *The Einstein Affair,* Orgone Institute Press, Rangeley,

Chapter References: 2. Timeline

Maine, 1953. see especially AE's letter of 7 Feb.1941 and Reich's response of 20 Feb.1941.

99. Cowley, Malcolm: "The End of the New Deal", *New Republic* 31 May 1943. p.729-732.

100. Bennett, Philip: "The Persecution of Dr. Wilhelm Reich by the Government of the United States", *International Forum of Psychoanalysis*, 19:51-65, 2010.

101. DeMeo, James: "Experimental Confirmation of the Reich Orgone Accumulator Thermal Anomaly", *Subtle Energies,* 20(3):1-16, 2010.

102. Reich, Wilhelm: *Discovery Of The Orgone, I: Function Of The Orgasm,* Orgone Institute Press, NY, 1942.

103. Published and edited by Reich in four volumes from 1942 through 1945. For full contents, see citation 6.

104. DeMeo, James: "Orgone Accumulator Stimulation of Sprouting Mung Beans", *Heretic's Notebook: Pulse of the Planet 5*:168-176 2002. Internet download: http://www.orgonelab.org/DeMeoSeedsSubtleEnergies.pdf

105. Meyerson, A.: "Book Reviews: *The Function of the Orgasm. Sex-Economic problems of Biological Energy ,*" *American Journal of Psychiatry*, 99:467-468, 1 November 1942. anon.: Distorted review of Reich's *Function of the Orgasm*, in *Journal of the American Medical Association,* 120(15):1260, 12 Dec. 1942.

106. This quote appears to be from a letter by Menninger to Fenichel, who then spreads the malicious lies throughout the psychoanalytic community in a 15 May 1942 round-letter. See the "Rundbrief 89" entry in Otto Fenichel: *Rundbrief 911*, Vol.2, page 1629 (See citation 54)

107. I.e., Wolfe, Theodore: "Misconceptions of Sex-Economy as Evidence in Book Reviews", *Int. J. Sex-Economy & Orgone Res.*, II:166--172, 1943. See citation 63 regarding the major publications attacking and slandering Reich, with response/reply articles.

108. Grotjahn, Martin: "Book Reviews: Recent Psychoanalytic Literature", *Psychosomatic Medicine,* 5, p.309-310, 1943.

109. These devices are described in: Reich, Wilhelm: *The Cancer Biopathy,* Orgone Institute Press, NY, 1948

110. Reich, Wilhelm: *Character Analysis: Principles and Technique for Psycho-analysts in Practice and in Training,* 2nd Enlarged Edition. English translation by Theodore Wolfe of 1933 edition (with new materials), Orgone Institute Press, New York, 1945.

111. Reich, Wilhelm: *The Sexual Revolution*, Orgone Institute Press, New York, 1945; 2nd Edition 1948.

112. Reich, Wilhelm: *The Mass Psychology of Fascism,* Orgone Institute Press, Rangeley, Maine, 1946.

113. For an excellent discussion on why Reich revised his American edition of the 1946 edition of his *Mass Psychology of Fascism,* see his own "Preface to the Third Edition". In the author's opinion, even in this revised edition Reich still speaks with ignorance about historical issues that would only gain public exposure in the coming decades. The massacres of the Stalinist gulag concentration labor-death camps were getting public discussion in the mid-1950s, but not to the extent as we know them today. The crimes of Lenin and Trotsky are still hardly known among left-intellectual circles, or they are

simply denied. The 100 million victims of rabid-dog Communism and the gulag in the 20th Century is as denied by the modern Left, as are the 11 million victims of Hitler's concentration death-camps denied by any neo-Nazi or Islamic Jew-hater one could find. In Reich's time, there was an information war going on no less than what we see today, where dominant leaders in government and news media were excessively friendly with "Uncle Joe" Stalin, who "helped us defeat Hitler" (after first being an ally of Hitler against the democratic West, and fully betraying Spain to the Nazis even before the Hitler-Stalin Pact, etc.).

Reich also appears certainly to have *not* known about the scheming hatred against human life and freedom as found in the concealed writings of Marx, Engels, Lenin and Trotsky. Those materials frequently contain open calls for holocaust massacre of "counter-revolutionaries" and "bourgeois capitalists", speaking crudely and derisively about "Jews, Poles, Slavs" and "niggers", whom they felt were incapable of "revolutionary thought" and who hence ought to be *wiped out*. Likewise, the actual conduct of Lenin and Trotsky, who began executing their opposition on the first days after shooting their way into power, was a long covered-up Soviet state secret, given how it would predictably offend pro-Soviet Western liberals. The Bolsheviks waged open war against civilians, and institutions such as the Cheka, NKVD and Red Army continued doing so under Soviet banners, committing many atrocities and war-crimes *well before* Stalin gained power. Regarding the cooperative schemes of the German High Command with Lenin during WW-I, and afterwards of the independent German military working secretly with Stalin, during Weimar even before Hitler, towards rearmament for yet another world war of global conquest (see citation 3 ibid.), much of that information is even today only slowly coming into the light of day. The political Left is in open denial about this as well. As noted in the text, Reich was put on an NKVD death list without his knowledge, along with one of his associates who was caught and "disappeared" into the maw of the gulag.

Reich's early view of "rational Marxist socialism against oppression" was largely a European illusion for which his American experience gave little validation. And so he revised his European writings to reflect the lessons learned. Americans had thrown both Kings and priests from power long before European Kings, Kaisers, Tsars and other dictators were deposed. Marxist historical analysis, as frequently written by Engels, carried an extreme bias against historical facts which suggested other pathways (such as democratic capitalism tempered by trade unions) towards social freedom outside of the violence-demanding steps outlined in the *Communist Manifesto*. In my opinion, Marx can easily be ignored in the body of Reich's work without loss of the essentials of his sex-economic or orgone biophysical discoveries, which actually owe a far greater debt to Freudian libido theory and classical biology. Also see:

- *Marx-Engels Genocide Quotes*: The Hidden History of Marx and Engels, Internet posted: http://www.orgonelab.org/MarxEngelsQuotes.htm
- Radosh, Ronald, Mary Habeck & Grigory Sevostianov, Editors: *Spain*

Chapter References: 2. Timeline

Betrayed: The Soviet Union in the Spanish Civil War, Annals of Communism Series, Yale Univ. Press, 2011.

114. Wertham, Fredric: "Calling All Couriers" (Book Review of The Mass Psychology of Fascism), *New Republic*, 2 December 1946.

115. *Annals of the Orgone Institute*, in two volumes from 1947 through 1948. See citation 6. The second volume of *Annals* is later reprinted as the book *Ether, God and Devil*. (See citation 132)

116. Brady, Mildred E.: "The New Cult of Sex and Anarchy", *Harpers Magazine*, April 1947, p.312-322. Brady, Mildred E.: "The Strange Case of Wilhelm Reich", *New Republic*, 26 May 1947; reprinted in *Bulletin of the Menninger Clinic*, 12(2):61-67, March 1948; condensation reprinted as "Is the World Sexually Sick?", *Everybody's Digest*, December 1947. For a discussion on the influential nature of *New Republic*, and on Michael Straight, see M.S. Evans and H. Romerstein, *Stalins Secret Agents: The Subversion of Roosevelt's Government*, Threshold Editions, NY 2012, p.137-138.

117. Martin, *Wilhelm Reich and the Cold War*, 2000, ibid, p.484-486.

118. Karl Menninger wrote 14 different short articles or book reviews in *The Nation* between 1937 and 1942, and at least 11 different items for *The New Republic* between 1938 to 1992. (See their on-line Archives.) One is forced to wonder why Menninger chose these radical-left publications for his articles, given how his popularity would easily have allowed many other possibilities.

119. Greenfield, Jerome: *Wilhelm Reich Vs. The USA*, W.W. Norton, NY, 1974.

120. Letter from Reich to Hays, 29 Nov.1947, with subsequent reply from Hays through his associate Julian Culver, in Reich, *American Odyssey*, ibid, p.427-429.

121. Reich, Wilhelm: *The Cancer Biopathy*, Orgone Institute Press, NY, 1948.

122. Reich, Wilhelm: *Listen, Little Man!*, Orgone Institute Press, NY, 1948.

123. Wolfe, Theodore P.: *Emotional Plague Versus Orgone Biophysics: The 1947 Campaign*, Orgone Institute Press, NY, 1948.

124. Greenfield, Jerome: "Examination of Wilhelm Reich by the Immigration and Naturalization Service", *Journal of Orgonomy*, 16(2):277-284, 1982.

125. *Orgone Energy Bulletin*, four issues per year, from 1949 to 1952. See citation 6.

126. Müschenich, Stefan & Gebauer, Rainer: *Die (Psycho-) Physiologischen Wirkungen des Reich'schen Orgonakkumulators auf den Menschlichen Organismus (The Psycho-Physiological Effects of the Reich Orgone Accumulator)*, Dissertation, Department of Psychology, University of Marburg, West Germany, 1986. Republished as: *Der Reichsche Orgonakkumulator: Naturwissenschaftliche Diskussion, Praktische Andwendung, Experimentelle Untersuchung*, Nexus Verlag, Frankfurt, 1987.

127. Hebenstreit, Günter: *Der Orgonakkumulator Nach Wilhelm Reich. Eine Experimentelle Untersuchung zur Spannungs-Ladungs-Formel*, Diplomarbeit zur Erlangung des Magistergrades der Philosophie an der Grund- und Integrativwissenschaftlichen Fakultät der Universität Wien, 1995.

128. anon.: "Cancer and the Need for Facts: Orgone Accumulator", Report of the Council on Pharmacy and Chemistry, *Journal of the American Medical Association*, 139(2):96-97, 8 January 1949.

Chapter References: 2. Timeline

129. anon.: "Cancer 'Cures' Beware", *Consumer's Reports,* March 1949. *Consumer's Reports* was originally formed by a communist outfit, the *Consumer's Union*, whose members had staged a violent strike and takeover of the original non-political product-testing organization, *Consumer's Research.* CU included Mildred Brady in its ranks, as well as other hard-left individuals.

130. Raphael, Chester M.: "Social Pathology: The Marlboro Incident", *Orgone Energy Bulletin*, I(2):70-76, 1949.

131. Gardner, Martin: "The Hermit Scientist", *Antioch Review*, winter 1950-1951.

132. Reich, Wilhelm: *Ether, God and Devil*, Orgone Institute Press, Maine, 1951.

133. Reich, Wilhelm: *Cosmic Superimposition: Man's Orgonotic Roots In Nature*, Wilhelm Reich Foundation, Maine, 1951.

134. Reich, Wilhelm: *The Orgone Energy Accumulator, Its Scientific And Medical Use*, Orgone Institute Press, Maine, 1951.

135. Reich, Wilhelm: The Oranur Experiment, First Report (1947-1951), Wilhelm Reich Foundation, Maine, 1951.

136. DeMeo, James: "Influences of the Reich Orgone Accumulator Upon Water, and By Water", *Sixth Annual Conference on the Physics, Chemistry and Biology of Water*, Mt. Snow, Vermont, October 2011.

137. Audiotape "Alone", April 3, 1952. Available from the Wilhelm Reich Museum bookstore. http://www.wilhelmreichtrust.org

138. Raphael, Chester & MacDonald, Helen E.: *Orgonomic Diagnosis of Cancer Biopathy,* Wilhelm Reich Foundation, Maine, 1952. Originally published as *Orgone Energy Bulletin*, IV(2):65-128, 1952.

139. Reich, *Wilhelm: Reich Speaks of Freud,* Interview with Kurt Eisler of the Freud Archives, Farrar, Straus & Giroux, NY, 1967.

140. Gardner, Martin: chapter on "Orgonomy" in *In The Name of Science* (later titled *Fads and Fallacies In the Name of Science*), Dover, NY 1952.

141. Reich, Wilhelm: *People In Trouble,* Orgone Institute Press, Maine, 1953.

142. Reich, Wilhelm: *The Murder of Christ*, Orgone Institute Press, Maine, 1953.

143. Reich, Wilhelm: "The Blackening Rocks: Melanor", *Orgone Energy Bulletin*, V(1-2):28-59, 1953.

144. *CORE* or *Cosmic Orgone Engineering*, published in four volumes over 1955-1956.

145. Reich, Wilhelm: "OROP Desert. Part 1: Spaceships, DOR and Drought", *Cosmic Orgone Engineering*, VI(1-4):1-140, 1954.

146. Reich, Wilhelm: *Contact With Space,* Oranur Second Report: Orop Desert Ea, Core Pilot Press, NY, 1957.

147. Reich, Wilhelm: *RESPONSE Regarding the Request of the Food and Drug Administration (FDA) to Enjoin the Natural Scientific Activities of Wilhelm Reich, M.D.,* to the Hon. Judge Clifford, Portland Maine, 25 Feb. 2954. Reproduced in *OROP Desert #2*, Orgone Institute Press, March 1954, p.4, and in *Response to Ignorance*, Orgone Institute Press, 1955, p.15-18.

148. Reich, Wilhelm: *Conspiracy: An Emotional Chain Reaction*, Orgone Institute Press, Rangeley, Maine, 1954.

149. *Orgonomic Medicine*, in three issues, 1955 to 1956, edited by Reich's

associate, Elsworth Baker.

150. DeMeo, James: *Preliminary Analysis of Changes in Kansas Weather Coincidental to Experimental Operations with a Reich Cloudbuster*, with the Appendix *Evidence for a Principle of Atmospheric Continuity*, and a New Preface. Reprint of author's 1979 research study from the Univ. of Kansas at Lawrence. Orgone Biophysical Research Lab, Ashland, Oregon, 2010.

151. DeMeo, James: *A 'New' Method for Drought-Abatement and Desert-Greening*. Internet Posting:
     http://www.orgonelab.org/ResearchSummary2.htm

152. Maglione, Roberto: *Wilhelm Reich and the Healing of Atmospheres*, Natural Energy Works, Ashland, Oregon, 2007.

153. Reich, Wilhelm: *Response To Ignorance*, Wilhelm Reich Biographical Material, History of the Discovery of the Life Energy, documentary supplement No. 1, A-XII-EP, Orgone Institute Press, Maine, 1955.

154. Reich, Wilhelm: *The Red Thread of a Conspiracy,* Wilhelm Reich Biographical Material, History of the Discovery of the Life Energy, Documentary Supplement No. 2, A-XII-EP, Orgone Institute Press, Maine, 1955.

155. Reich, Wilhelm: *Atoms for Peace Vs. the HIG* (Hooligans In Government), Wilhelm Reich Biographical Material, History of the Discovery of the Life Energy, Documentary Supplement No. 3, A-XII-EP, Orgone Institute Press, Maine, 1956.

156. Ruppelt, Edward J.: *Report on Unidentified Flying Objects*, Ace Books/ Doubleday, NY 1956, see Chapter 5, "The Radiation Story", p.262-274.

157. Reich, W.: *Record Appendix to Briefs for Appellants, Vol.III: Suppressed Documentary Evidence,* U.S. Court of Appeals, for the First Circuit, No. 5160, Wilhelm Reich, et al., Defendants-Appellants, v. United States of America, Appellee, 1957; LC# 57-8813.

158. Reich, W.: *Record Appendix to Briefs for Appellants, Vol.V: Suppressed, Top Secret Evidence, OROP Desert Ea,* U.S. Court of Appeals, for the First Circuit, No. 5160, Wilhelm Reich, et al., Defendants-Appellants, v. United States of America, Appellee (Prepared but never published or released).

159. Baker, E.F., et al.,*Pettion for a Writ of Certiorari to the US Court of Appeals for the First Circuit,* Supreme Court of the United States, October Term 1955, Elsworth F. Baker, et al., Petitioners, v. U.S.A., Respondent.

160. *Wilhelm Reich et al., Defendants, Appellants, v. United States of America, Appellee. United States Court of Appeals First Circuit.* - 239 F.2d 134. Hearing November 6, 1956. Decided December 11, 1956. Writ of Certiorari Denied February 25, 1957. See specifically Points #17 through #20, and footnote 1 in particular, which documents that the Brady slanders from the *New Republic* had entered into the court case record, as if they were factual material, and additionally how the Supreme Court agreed with the FDA Injunction and lower court rulings *demanding the burning of books!* See:
     http://law.justia.com/cases/federal/appellate-courts/F2/239/134/108394/
     This above document also referenced (in Point 18) in the determination against Reich, a prior opinion by Justice Frankfurter to uphold a Contempt of Court conviction against the United Mine Workers, as given on page 312:
     http://supreme.justia.com/cases/federal/us/330/258/case.html

Frankfurter's high and mighty words, from an older ruling on a major labor dispute, were dusted off and used against Reich by the Supreme Court to justify their decision to ignore his documentation on FDA *lying and fraud* before the lower courts, and to basically *declare in favor of the burning of books.*

Frankfurter: *"In our country, law is not a body of technicalities in the keeping of specialists or in the service of any special interest. There can be no free society without law administered through an independent judiciary. If one man can be allowed to determine for himself what is law, every man can. That means first chaos, then tyranny. ... In a democracy, power implies responsibility. The greater the power that defies law, the less tolerant can this Court be of defiance. As the Nation's ultimate judicial tribunal, this Court, beyond any other organ of society, is the trustee of law and charged with the duty of securing obedience to it."*

161. anon.: "Book Order Appealed: Liberties Unit Asks U.S. Not to Destroy Reich's Writings", *New York Times*, 13 July 1956.

162. Matusow, Harvey: *False Witness*, Cameron & Kahn, NY 1955. I met Matusow in Lawrence, Kansas around 1978. He was giving a "bell ringing" concert and when he discovered I was doing research on the Reich cloud-buster, he approached me with the story of his being in the prison cell next to Reich. He blamed "McCarthyism" for Reich's troubles and death in prison, and openly stated he had worked to destroy McCarthy, by deliberately introducing lies into Congressional testimony. At the time I had no idea who he was or the significance of his statement. Matusow's book carried a Foreword by Albert E. Kahn, a Soviet sympathizer and spy, and author of books attacking both the HUAC and FBI efforts to oppose communism.

163. Haynes, John Earl: *Senator Joseph McCarthy's Lists and Venona.* Haynes re-compiled the original McCarthy lists of accused communists and compared them to what is known today from Soviet decryptions, archives and other new information. Haynes wrote: *"Of the 159 persons listed ... there is substantial evidence that nine assisted Soviet espionage against the United States... Some of the others were security risks. ... Risks should be minimized by excluding those persons from employment in positions where they would have access to sensitive information. ... My own view is that a number of those on the lists... perhaps a majority, likely were security risks..."*
    http://www.johnearlhaynes.org/page62.html
    Also see citation 95 and Evans, Stanton M.: *Blacklisted by History: The Untold Story of Senator Joe McCarthy and His Fight Against America's Enemies,* Three Rivers Press, 2009.

164. McCarthy personally described Murrow's communist associations in a television rebuttal: "Senator Joseph R. McCarthy: Reply to Edward R. Murrow", *See it Now*, CBS-TV broadcast, April 6, 1954. Internet posted:
    http://www.lib.berkeley.edu/MRC/murrowmccarthy2.html
    McCarthy exposed Murrow's pro-Soviet broadcasting and, among other things, revealed Murrow having a book dedicated to him by British communist Harold Laski (*Reflections on the Revolution of Our Times* c.1943). Laski also wrote the Foreword to Robert Brady's book declaring American capitalism to be equal to Nazism (see citation 88). Laski was also a "good friend" of the pro-Soviet Felix Frankfurter (see citation 94) who sat on the US Supreme

Chapter References: 3. After Reich

Court, ruling against Reich and in favor of book-burning. Reich's detractors were intimately connected by a web of shared communist, pro-Soviet agendas and friendships, working against anyone with anti-Communist or anti-Stalinist ideas, or who merely held American liberty and the US Constitution in high esteem.

165. Matusow was interviewed by German filmmaker Digne Marcovicz for her production of *Viva Little Man*. http://www.vivalittleman.com   Also see: Matusow, Harvey: "The Death of Wilhelm Reich", Internet posted: http://www.ibiblio.org/mal/MO/matusow/chapters/ch14.html

Chapter 3: After Reich's Death

1. *Wilhelm Reich Infant Trust Fund,* and *Wilhelm Reich Museum*, Rangeley, Maine, USA. http://www.wilhelmreichtrust.org
2. Ritter, Paul: "The Persecution of Wilhelm Reich and the Silence of the Press", *Orgonomic Functionalism*, IV(1):3-14, 1957.
3. Ritter, Paul (editor): *Wilhelm Reich Memorial Volume* , Ritter Press, Nottingham, England 1958.
4. Boadella, David: "The Murder of Reich, 1957", *Orgonomic Functionalism*, IV(6):250-253, 1957.
5. Wyvell, Lois & Rees, Raymond: "The Jailing of a Great Scientist in the USA, 1956", privately published, NY, 1956.
6. Kelley, Charles: *The Ending of Wilhelm Reich's Researches*, Interscience Research Institute, Stamford, CT, 1960.
7. Reich, Wilhelm: *Selected Writings: An Introduction to Orgonomy*, Farrar, Straus & Giroux, NY, 1960
8. Please review the list of Reich's major books given in the Introduction Chapter.
9. The ACO is a professional association and not a teaching college in the usual sense. They publish the *Journal of Orgonomy*.   http://www.orgonomy.org
10. The *Annals* is published by the Institute for Orgonomic Science. http://www.orgonomicscience.org
11. *Pulse of the Planet*, published by the Orgone Biophysical Research Lab. http://www.orgonelab.org/cart/xpulse.htm
12. For a full review, see the online *Bibliography on Orgonomy* http://www.orgonelab.org/bibliog.htm
13. See the online *Emotional Plague Bibliography* for a more complete listing of these attack-smear publications: http://www.orgonelab.org/bibliogPLAGUE.htm
14. Makavejev, Dusan: *WR Mysteries of the Organism*, Film Release, 1971.
15. Crombie, Albert: *Communism and the Moral Breakdown in America*. Clearwater, FL: Youth Problems, 1955
16. Flynn, Daniel: *A Conservative History of the American Left*, Crown Forum 2008.
17. Bell, John: "An Attack by the Radical Left: A Report", *Journal of Orgonomy*, 8(1):65-78, 1974. Montgomery, L.: "5 Classes Taken From Teacher; Action Sparks Student Protest", *Miami Herald*, 29 April 1975, p.1,4-B.

18. DeMeo, James: *Preliminary Analysis of Changes in Kansas Weather Coincidental to Experimental Operations with a Reich Cloudbuster*, Republication of author's 1979 research study from the University of Kansas at Lawrence. Orgone Biophysical Research Lab, Ashland, Oregon, 2010.

19. DeMeo, James: Lecture "Field Experiments with the Cloudbuster: 1977-1983", Geography-Geology Department, Illinois State University, 15 Sept.1983.

20. DeMeo, James: "Evidence for the Existence of a Principle of Atmospheric Continuity", contained in DeMeo citation 18, ibid.

21. DeMeo, James: On the Origins and Diffusion of Patrism, the Saharasian Connection. Geography Department, University of Kansas, Lawrence 1986. Republished as: *Saharasia: The 4000 BCE Origins of Child-Abuse, Sex-Repression, Warfare and Social Violence, In the Deserts of the Old World*, Revised and Updated Second Edition, Natural Energy Works, Ashland, Oregon 2006.   http://www.saharasia.org

22. The central most "skeptic group" has been CSICOP – *Committee for Scientific Investigation of Claims of the Paranormal*, which publishes *Skeptical Inquirer* magazine. At the height of its popularity, CSICOP had dozens of affiliated "clubs" scattered around the USA and in Europe. Their advisory board included dozens of high-ranking professors and editors of major science journals, as well as some darkly perverse types. See the Appendix for documentation on that. For specific details on attacks against myself and other Reich scholars, see the following:
    - Martin, Jim & Thomas, Kenn: "Toxic Disinformation: Joel Carlinsky's Bonfire of Insanity", *Steamshovel Press,* 1998. Internet Posted:
        http://www.umsl.edu/~skthoma/carl.htm
    - DeMeo, James: *"Open Letter on Joel Carlinsky and the Organized 'Skeptics' War Against American Orgonomy"*, 2005. Internet Posted:
        http://www.orgonelab.org/carlinskyletter.htm

23. DeMeo, James: "Response to Martin Gardner's Attack on Reich and Orgone Research in the *Skeptical Inquirer*", *Pulse of the Planet* 1:11-17 Spring 1989. http://www.orgonelab.org/gardner.htm

24. The Orgone Biophysical Research Lab, Ashland, Oregon, USA http://www.orgonelab.org

25. Reich, Wilhelm: "Emotional Plague" *Selected Writings,* Farrar, Straus & Giroux, 1973, p.467-513; W. Reich, *Character Analysis*, Farrar, Straus & Giroux, 1961, p.504-539. W. Reich, *The Murder of Christ*, Farrar, Straus & Giroux, 1971.

26. DeMeo, James, et al (with 23 coauthors): "In Defense of Wilhelm Reich: An Open Response to Nature and the Scientific/Medical Community," *Water: A Multidisciplinary Research Journal*, V.4, p.72-81, 2012. Internet posting: http://www.waterjournal.org/volume-4

27. DeMeo, James: *Maps From the Ethnographic Atlas Data: A Defense of the Cross-Cultural Codes and Data Base of G. P. Murdock and the Quadruple-Blind Control Procedures Used in my Saharasia Research*, 2007. Internet posting: http://www.orgonelab.org/murdockdata.htm

28. Prescott, James: "Body Pleasure and the Origins of Violence", *Pulse of the Planet* 3:17-25 1991.

29. Baker, Elsworth F.: *Man In The Trap*, Macmillan, NY, 1967.
30. Herskowitz, Morton: *Emotional Armoring: An Introduction to Psychiatric Orgone Therapy*, Transactions Press, NY 1998.
31. Müschenich, Stefan: *Der Gesundheitsbegriff im Werk des Arztes Wilhelm Reich (The Concept of Health in the Works of Dr. Wilhelm Reich)*, Verlag Gorich & Weiershauser, Marburg 1995.
32. Reich, Wilhelm: *The Bioelectrical Investigation of Sexuality and Anxiety*, Farrar, Straus & Giroux, NY, 1982.
33. Burr, H.S.: *Blueprint For Immortality,* Neville Spearman, London, 1971; *The Fields of Life,* Ballantine Books, NY, 1972.
34. Becker, R.O. & Selden, G.: *The Body Electric: Electromagnetism and the Foundation of Life,* Wm. Morrow, NY, 1985.
35. Nordenstrom, B.: *Biologically Closed Electric Circuits:,* Nordic Medical Press, Stockholm, 1983.
36. Braid, Byron & Dew, Robert: "Reich's Bioelectric Experiments: A Review With Recent Data", Annals, Institute for Orgonomic Science, 5(1):1-18, 1988.
37. Hebenstreit G.A.: *Die Elektrophysiologie erogener und nicht-erogener Hautzonen in Verbindung mit der Spannungs-Ladungs-Formel Wilhelm Reichs. Eine experimentelle Untersuchung.* Universität Wien, Dissertation, 2010. http://othes.univie.ac.at/13854/1/2010-09-08_8501559.pdf
38. Reich, Wilhelm: *The Bion Experiments: On the Origin of Life,* Farrar, Straus & Giroux, NY, 1979. Also see "Laboratory Manual for Bion Experiments: Notes by a Laboratory Worker" Norway, February 1938 (Handwritten & Typed Pages in German only), Photocopy, Wilhelm Reich Museum, Rangeley.
39. DeMeo, James: "Bion-Biogenesis Research and Seminars at OBRL: Progress Report", *Heretic's Notebook: Pulse of the Planet* 5:100-113 2002.
40. Reich, Wilhelm: *The Cancer Biopathy, Vol.2, Discovery of the Orgone*, A. White, trsl., Farrar, Straus & Giroux, NY, 1973.
41. Müschenich, Stefan & Gebauer, Rainer: *Der Reichsche Orgonakkumulator: Naturwissenschaftliche Diskussion, Praktische Andwendung, Experimentelle Untersuchung*, Nexus Verlag, Frankfurt, 1987.
42. Hebenstreit, Günter: *Der Orgonakkumulator Nach Wilhelm Reich. Eine Experimentelle Untersuchung zur Spannungs-Ladungs-Formel*, Diplomarbeit zur Erlangung des Magistergrades der Philosophie an der Grund- und Integrativwissenschaftlichen Fakultät der Universität Wien, 1995.
43. DeMeo, James: *The Orgone Accumulator Handbook: Wilhelm Reich's Life-Energy Science and Healing Tools for the 21st Century, With Construction Plans*, 3rd Revised Edition, Natural Energy Works, Ashland, Oregon, 2010.
44. Kavouras, Jorgos: *Heilen mit Orgonenergie: Die medizinische Orgonomie*, Turm Verlag, Bietigheim, Germany, 2005.
45. Friedlander, Henry: *The Origins of Nazi Genocide: From Euthanasia to the Final Solution*, Univ. North Carolina Press, 1997.
46. DeMeo, James: "Anti-Constitutional Activities and Abuse of Police Power by the U.S. Food and Drug Administration and other Federal Agencies," *On Wilhelm Reich and Orgonomy, Pulse of the Planet* #4:106-113, 1993. Walker, Martin: *Dirty Medicine: Science, Big Business and the Assault on Natural Health Care*, Slingshot Publications, London 1993. Carter, James P.:

Chapter References: 3. After Reich

Racketeering In Medicine: The Suppression of Alternatives, Hampton Roads, Oakland 1992.

47. Senf, Bernd: "Wilhelm Reich, Discoverer of Acupuncture Energy", *Pulse of the Planet* 2:25-30 Fall 1989; Southgate, Leon: *Chinese Medicine and Wilhelm Reich: An Analysis of Chinese Medical and Reichian Theories of Life Force And Experimental Orgone-Acupuncture Study*, Lambert Academic Publishing, London 2009.

48. Mazzocchi A, Maglione R : "A Preliminary Study of the Reich Orgone Accumulator Effects on Human Physiology", *Subtle Energies and Energy Medicine.* 21(2):41-50, 2010.

49. DeMeo, James: "Report on Orgone Accumulator Stimulation of Sprouting Mung Beans", *Subtle Energies and Energy Medicine*, 21(2):51-62, 2010. http://www.orgonelab.org/DeMeoSeedsSubtleEnergies.pdf

50. *Optimal Orgonotic Conditions*: In addition to other standard control procedures, proper orgone accumulator experiments require low humidity (<50%), air clear of significant chemical pollutants, a generally higher altitude above any atmospheric haze-layer, preferably within forest regions, an effective distance (20-50 miles) from nuclear power plants, as well as a laboratory environment free from stray low-level electromagnetic fields ("electrosmog") as from fluorescent lights, computers, CRT devices, WiFi equipment, cell-phone tower RF, and so on. An orgone accumulator may work well in spite of some of these factors, but if it does not, then for more exacting work one must default to the more strict optimal initial conditions for valid results. This kind of laboratory situation was characteristic of Reich's *Orgonon* facility, and is also true of DeMeo's *OBRL-Greensprings* facility.

51. DeMeo, James: "Water as a Resonant Medium for Unusual External Environmental Factors", *Water: A Multidisciplinary Research Journal*, V.3, p.1-47, 2011.

52. Maglione, Roberto: *Methods and Procedures in Biophysical Orgonometry*, Moncrivello, Italy, 2012.

53. DeMeo, James: "Experimental Confirmation of the Reich Orgone Accumulator Thermal Anomaly", *Subtle Energies and Energy Medicine*, 20(3):1-16, 2009.

54. DeMeo, James: "Influences of the Reich Orgone Accumulator Upon Water, and By Water", *Sixth Annual Conference on the Physics, Chemistry and Biology of Water*, Mt. Snow, Vermont, October 2011.

55. DeMeo, James: "Field Experiments with the Reich Cloudbuster: 1977-1983", *Journal of Orgonomy*, 19(1):57-79, 1985; DeMeo, James & Robert Morris: "Preliminary Report on a Cloudbusting Experiment in the Southeastern Drought Zone, August 1986", *Southeastern Drought Symposium Proceedings,* March 4-5, 1987, South Carolina State Climatology Office Publication G-30, Columbia, SC, 1987; DeMeo, James: "CORE Progress Report #26: California Drought of 1990-1991, Part II. With a Special Note on Underground Nuclear Testing and the Oakland Wildfires", *Journal of Orgonomy*, 26(1):,49-71 1992; parts reprinted in: "The Oakland Wildfires of October 1991", *Pulse of the Planet* 4:117-118,1993.

56. DeMeo, James: "OROP Arizona 1989: A Cloudbusting Experiment to Bring

Chapter References: 4. Review of Turner

Rains in the Desert Southwest", *Pulse of the Planet*, 3:82-92, 1991.

57. DeMeo, James: "OROP Israel 1991-1992: A Cloudbusting Experiment to Restore Wintertime Rains to Israel and the Eastern Mediterranean During an Extended Period of Drought", *Pulse of the Planet* 4:92-98, 1993.

58. DeMeo, James: "Research Reports and Observations: OROP Namibia 1992-1993", *Pulse of the Planet* 4:114-116, 1993.

59. DeMeo, James: "Green Sea Eritrea: A 5-Year Desert Greening CORE Project in the SE African Sahel", *Pulse of the Planet*, 5:183-211, 2002.

60. Maglione, Roberto: *Wilhelm Reich and the Healing of Atmospheres*, Natural Energy Works, Ashland, Oregon, 2007.

61. Sharaf, Myron: *Fury on Earth: A Biography of Wilhelm Reich*, St. Martin's-Marek, NY, 1983. For a good summary of Reich's work in Sex-Pol, see Chapters 4 and 5:

62. Boadella, David: *Wilhelm Reich: The Evolution of His Work,* Henry Regenry, Chicago, 1973.

63. Raknes, Ola: *Wilhelm Reich and Orgonomy*, St. Martin's Press, NY, 1970

64. Greenfield, Jerome: *Wilhelm Reich Vs. The USA*, W.W. Norton, NY, 1974.

65. Martin, Jim: *Wilhelm Reich and the Cold War*, Flatland Books, Mendocino, CA, 2000. Revised edition, Natural Energy Works, Ashland, OR, 2013.

66. Ollendorff, Ilse: *Wilhelm Reich: A Personal Biography*, St. Martin's, NY, 1969.

67. DeMeo, James & Senf, Bernd (Eds.): *Nach Reich: Neue Forschungen zur Orgonomie: Sexualökonomie , Die Entdeckung der Orgonenergie (After Reich: New Research in Orgonomy: Sex-Economy, Discovery of the Orgone Energy)*, Zweitausendeins Verlag, Frankfurt, 1997.

68. DeMeo, James (Ed.): *Heretic's Notebook: Emotions, Protocells, Ether-Drift and Cosmic Life Energy, With New Research Supporting Wilhelm Reich* (Pulse of the Planet #5) Natural Energy Works, Ashland, Oregon 2002.

69. DeMeo, James (Ed.): *On Wilhelm Reich And Orgonomy* (Pulse of the Planet #4), Natural Energy Works, Ashland, Oregon 1993.

70. Boadella, David, ed.: *In the Wake of Reich*, Coventure, London, 1976.

71. Hinchey, Kevin, Writer and Director: *Man's Right to Know.* Privately produced. Rangeley Maine 2002.

72. East, Jon, Writer and Director: *It Can Be Done*, Privately Produced, London 1999.

73. Marcovicz, Digne, Director: *Viva Little Man, Privately* Produced, Berlin Germany 1987.

74. Svoboda, Antonin, Writer and Director: *Whose Afraid of Wilhelm Reich*, Privately Produced, Austria 2009.

75. Exceptions here are Dusan Makavejev's film *WR Mysteries of the Organism,* reviewed in Chapter 6, and Antonin Svoboda's 2nd film on Reich from 2012, *The Strange Case of Wilhelm Reich.* The latter film of Svoboda creates a fictional Reich who is misportrayed as a victim of American right-wing extremists, without a single reference to the many leftists, communists and even Soviet agents who figured prominently in his death. The film also fabricates the claim that Reich's last wife, Aurora Karrer, was a *CIA Agent* out to destroy Reich. The very real Soviet spies and KGB agents who put

Reich on a death-list in Europe, and later worked to destroy Reich in America are never mentioned. But a claimed CIA agent is fabricated, sleeping in Reich's bed!

Chapter 4: Critical Review of Turner's *Orgasmatron*

1. http://www.wilhelmreichtrust.org
2. Turner, C., *London Review of Books*, 26(11), 3 June 2004, p.38-39. Also see: Turner, C.: "Free-for-All: A. S. Neill and Summerhill", Cabinet. #39, Fall 2010.
3. Wilmers, Mary Kay: *The Eitingons*, Verso Press 2010.
4. A.P. Watt.   http://www.apwatt.co.uk
5. http://www.londonconsortium.com/about/phdtitles.php
6. Reich, W.: *Bioelectrical Investigation of Sexuality and Anxiety*, Farrar, Straus & Giroux, NY 1982. Reprint of Reich's research reports from c.1937.
7. Reich, W.: *The Bion Experiments: On the Origins of Life,* Farrar, Straus & Giroux, NY 1979. Reprint of Reich's *Die Bione: Zur Entstehung des vegetativen Lebens,* Issued in the series *Klinische und Experimentelle Berichte (Clinical and Experimental Report), Institut für Sexual konomische Lebensforschung,* Sexpol Verlag, Oslo/Kopenhagen/Zürich, 1938.
8. Reich, Wilhelm: "The Discovery of the Orgone: Experimental Investigations of Biological Energy", *International Journal of Sex-Economy & Orgone Research,* I(2):108-130, July 1942. Reich, Wilhelm: *The Orgone Energy Accumulator, Its Scientific And Medical Use*, Orgone Inst. Press, Maine 1951.
9. Brady, Mildred E.: "The New Cult of Sex and Anarchy", *Harper's Magazine*, April 1947. Brady, Mildred E.: "The Strange Case of Wilhelm Reich", *New Republic,* 26 May 1947; reprinted in *Bulletin of the Menninger Clinic,* 12(2):61-67, March 1948.
10. Gardner, Martin: "The Hermit Scientist", *Antioch Rev,,* winter 1950-1951. Also see Gardner's chapter on "Orgonomy" in his book *In The Name of Science* (later titled *Fads and Fallacies In the Name of Science*), Dover, NY 1952.
11. Reich, Wilhelm: *Discovery Of The Orgone, II: The Cancer Biopathy,* T. Wolfe, trnsl., Orgone Institute Press, NY, 1948. Republished by Farrar, Straus & Giroux, NY 1973,
12. DeMeo, James: *The Orgone Accumulator Handbook: Wilhelm Reich's Life-Energy Science and Healing Tools for the 21st Century, With Construction Plans,* 3rd Revised Edition, Natural Energy Works, Ashland, Oregon, 2010.
13. Reich, W.: "Notes: The Orgone Accumulator and Orgastic Potency", *Orgone Energy Bulletin,* 2(2):93,1950.
14. Reich, W: "On the Record: Clarifications", *Orgone Energy Bulletin,* 4(4):218, 1952.
15. Reich, Wilhelm: *Atoms For Peace Vs. The HIG,* Orgone Institute Press, Rangeley, 1956 pp.9, 13, 24.
16. Wertham, Fredrick: "Calling All Couriers" (Book Review of *The Mass Psychology of Fascism*), *New Republic,* 2 December 1946.
17. Martin, J.: *Wilhelm Reich and the Cold War*, Flatland Press, Mendocino,

Chapter References: 4. Review of Turner

Calif. 2000. Revised edition, Natural Energy Works, Ashland, OR, 2013.

18. Reich, W.: *Function of the Orgasm*, Orgone Institute Press, 1942, p.360. Republished by Farrar, Straus & Giroux, NY 1973,

19. Reich, W.: *The Einstein Affair*, Orgone Institute Press, 1953. Specifically see the Letter from Albert Einstein to Wilhelm Reich, 7 Feb.1941, and Reich's reply to Einstein of 20 Feb.1941

20. *Journal of Orgonomy*, published by the American College of Orgonomy. http://www.orgonomy.org

21. DeMeo, James: "Experimental Confirmation of the Reich Orgone Accumulator Thermal Anomaly", *Subtle Energies,* 20(3):1-16, 2009.

22. Kinsey, Alfred: *Sexual Behavior in the Human Male,* W.B. Saunders, Philadelphia 1948.

23. My copy of the 1948 first edition of *SBHM* , reports in the front matter a total of 8 print-runs between January to August alone. http://www.kinseyinstitute.org/about/earlycontroversy.html

24. Kinsey, Alfred: *Sexual Behavior in the Human Female,* W.B. Saunders, Philadelphia 1953.

25. "Two Year Report of the Orgone Institute Press, 1950-1951", *Orgone Energy Bulletin*, IV(3):157-161, 1952.

26. Reisman, Judith, et al: *Kinsey, Sex and Fraud,* Vital Issues Press, 1990. Reisman, Judith: *Kinsey: Crimes and Consequences,* Inst for Media Education, 2003. http://www.drjudithreisman.com

27. "Kinsey's Paedophiles", Yorkshire TV production for Channel 4, aired 10 August 1998. On YouTube in 6 parts: http://www.youtube.com/watch?v=htAUysRPvNs

28. Jones, James: *Alfred C. Kinsey: A Public/Private Life,* W.W. Norton, 1997. Especially see pp. 619-626.

29. Kinsey 1948, ibid. p.259-261.

30. Gates, Gary: "How many people are lesbian, gay, bisexual, and transgender?" Williams Institute, 2011. Chandra, Anjani, et al.: "Sexual Behavior, Sexual Attraction, and Sexual Identity in the United States", National Health Statistics Report, #36,p.28-30, 2011.

31. Kinsey 1948, ibid. p.261-262.

32. Kinsey 1948, ibid. p.176-180

33. Kinsey 1948, ibid. p.161. See the section on "Extreme tension with violent convulsion".

34. Turner, C.: "Hugh Hefner in six volumes", *Guardian,* 17 July 2010.

35. Amanda Bailey and Randolph Trumbach. "Welcome to the Molly-House: An Interview with Randolph Trumbach", *Cabinet* #8. Fall 2002

36. Turner, C.: "Diary", *London Review of Books,* V.27 No.1, 6 Jan. 2005.

37. Kramer, Peter: "The Great Proselytizer of Orgasm: A new book explores the highly peculiar legacy of Wilhelm Reich", *Slate.com,* 27 June 2011.

38. Kramer, Peter: *Listening to Prozac,* Viking Press 1993.

39. Thomas, June: "The Gay Bar: Is it dying?" and "The Gay Bar - first gay-bar experience." *Slate.com*, 27 June 2011.

40. Reports on Mr. Savage's antics litter the internet. For example: http://www.wnd.com/2012/05/dan-savage-tolerant-bully/

http://www.wnd.com/2012/10/every-dead-gay-kid-is-a-victory-for-frc/
http://www.breitbart.com/Big-Government/2012/04/29/Savage-Republicans-Dead
http://www.rpvnetwork.org/profiles/blogs/dan-savage-we-should-acknowledge-the-existence-of-good-pedophiles
http://www.thestranger.com/seattle/Content?oid=3092
http://www.thestranger.com/seattle/SavageLove?oid=3347526

41. Sharaf, Myron: *Fury on Earth: A Biography of Wilhelm Reich,* St. Martin's Marek, NY, 1983, p.162-163.

42. Reich, W. "On Homosexuality", in *Children of the Future,* Farrar, Straus & Giroux 1983, p.200-201.

43. Kinsey 1948, ibid, p.159-160.

44. Reich *Function of the Orgasm,* ibid, p.96.

45. Hodann, Max: *History of Modern Morals,* Wm. Heinemann, London, 1937, p.57. Also see: "Notes: Jerry's Case" *Annals of the Orgone Institute,* 1:133-134, 1947.

46. Reich, W.: *Mass Psychology of Fascism,* Sex-Pol Verlag, Copenhagen 1933; Expanded 2nd Edition, Orgone Institute Press, Rangeley 1946.

47. Reich, Wilhelm: *Der Sexuelle Kampf der Jugend (The Sexual Struggle of Youth),* Verlag für Sexualpolitik, Berlin 1932. Reprinted as "The Sexual Rights of Youth" in Reich, *Children of the Future,* ibid, 1983.

48. Reich, Wilhelm: *Contact With Space, Oranur Second Report: Orop Desert Ea,* Core Pilot Press, NY, 1957

49. anon. "Morals: The Second Sexual Revolution", *Time,* 24 January 1964, p.54-59.

50. anon.: Article attacking Reich, *Time,* 14 April 1947. Also see the Emotional Plague bibliography: http://www.orgonelab.org/bibliogPLAGUE.htm

51. Frederic Golden, Leon Jaroff, Jeffrey Kluger and Michael D. Lemonick: "Cranks, Villains and Unsung Heroes" *Time ,* 29 March 1999, p.64-69 & 196.

52. See: Wilder, John: "CSICOP, Time Magazine, and Wilhelm Reich", *Pulse of the Planet* 5:55-66 2002.

53. A simple internet search on the phrases "Berlin Love Parade" or "Folsom Street Fair" will bring up hundreds of websites, most of which are sanitized so as to remove the violent, sadist and pedophile materials. However, if you add those latter terms into the search – "sadist", "pedophile", "urination", etc. – you will get a mix of shocking items sometimes with photographs, being either critical of such behavior, or outright defending of it and promoting it.

54. DeMeo, James: "Editor's Postscript: CSICOP, Prometheus Books, Pornography and the *Journal of Pedophilia*?!", *Heretic's Notebook: Pulse of the Planet* 5:65, 2002.

55. Malinowski, Bronislaw: War and Weapons Among the Natives of the Trobriand Islands. *Man* Vol. 20, (Jan., 1920), pp. 10-12.

56. DeMeo, James: *Saharasia: The 4000 BCE Origins of Child-Abuse, Sex-Repression, Warfare and Social Violence, In the Deserts of the Old World,* Revised and Updated Second Edition, Natural Energy Works, Ashland, Oregon 2006.

57. In November 1940, one year after his arrival in America, Reich wrote to his

friend A.S. Neill: *"...this so-called bourgeois society has done more in the course of seven years under capitalistic rules in the field of social security than any communist in Russia would dream of getting. ... If you hear from socialists and communists coming over here and claiming that Roosevelt is a dictator or a fascist, then your stomach simply turns around. I started to hate them."* (Reich, Wilhelm: *American Odyssey: Letters and Journals 1940-1947*, Farrar, Straus & Giroux, NY 1999, p.38.) Also see: Swarowsky, Anthony I.: "Thoughts on the Sex Behavior of American Soldiers in the ETO (European Theater of Operations)", *Annals of the Orgone Institute*, I:101-107, 1947.

58. Revel, Jean-Francois: *Without Marx or Jesus*, Doubleday 1971. Revel, Jean-Francois: *Anti-Americanism*, Encounter Books 2004.

59. Cott, Alan: "A Warning", *Orgone Energy Bulletin*, 1(1):35-36, 1949.

60. Ollendorff, Ilse: Wilhelm Reich, A Personal Biography, 1969, p.36.

61. Fenichel, Otto: *119 Rundbriefe: 1934-1945*, Stroemfeld, Frankfurt 1998.

62. W. Reich, *Conspiracy: An Emotional Chain Reaction*, Orgone Institute Press, 1954, Documents #157-161& #178 for 1948, p.12, 14.

63. anon.: "Cancer and the Need for Facts: Orgone Accumulator", Report of the Council on Pharmacy and Chemistry, *Journal of the American Medical Association*, 139(2):96-97, 8 January 1949.

64. anon.: "Cancer 'Cures' Beware", *Consumer's Report*, March 1949.

65. See the Emotional Plague bibliography:
     http://www.orgonelab.org/bibliogPLAGUE.htm

66. Flynn, Daniel: *A Conservative History of the American Left*, Crown Forum, NY, 2008. Crombie, Albert: *Communism and the Moral Breakdown in America*. Clearwater, FL: Youth Problems, 1955. Both of these books written by conservatives attacked Reich primarily from the standpoint of the pro-Soviet Brady smears of 1947.

67. Reich, Eva: "Orgonomic First Aid for Mothers and Infants", *Heretic's Notebook: Pulse of the Planet* 5:5-14 2002.

68. As quoted in Turner, 2011, p.321.

69. Baker 1990, As quoted in Turner, 2011, p.321.

70. Personal communication. Richard A. Blasband, 2011.

71. There is no evidence whatsoever – zero – that Reich was either advocating or involved in any kind of child abuse, sexual or otherwise. The only open question remaining is, if one or more of his trainees may have committed such assaults against children without Reich's knowledge or in the years after his death. It remains difficult to gain clarity on this latter issue, however, in a social atmosphere of a witch-hunt being conducted by a small army of Reich-haters, who only recently added the malicious slander of "child sexual abuse" to their long lists of false accusations.

72. Lowen, Alexander: "Review: The Ethics of Sexual Acts by Rene Guyon", *International Journal of Sex-Economy & Orgone Research*, III(2-3):205-210, October 1944.

73. Wyvell, Lois: "Orgone and You", *Offshoots of Orgonomy*, Issue #7, Autumn 1983, p.9.

74. Siersted, Ellen: "Wilhelm Reich in Denmark", *On Wilhelm Reich and Orgonomy: Pulse of the Planet* 4:44-69 1993.

Chapter References: 5. *Orgasmatron* Aftermath

75. On-Line Bibliography On Orgonomy: http://www.orgonelab.org/bibliog.htm
76. Aurora Karrer Archive, National Institute of Health, Bethesda, MD.
77. Ollendorff, Ilse: *Wilhelm Reich: A Personal Biography,* St. Martin's, NY, 1969, p.119.
78. "Report on injury to WR's dog Troll", Personal Files, Box 11, Reich Archive, Harvard Countway Library, Boston. Via personal communication from Kevin Hinchey of the Reich Trust, 7 May 2013. The file includes:
    1) Handwritten report from Eva Reich to the local Sheriff in Farmington, Maine, 7 September 1955. 2) Handwritten Letter from E. E. Russell, Veterinarian, Farmington Maine, 10 September 1955. 3) Letter from the State of Maine - Department of Health and Welfare, sent to Earl R. Hawkens, Sheriff, Franklin County, Farmington Maine, 22 September 1955.
79. *"Garbage was repeatedly strewn on Reich's lawn; air was let out of his tires; he received repeated telephone calls from anonymous people making remarks about the accumulator and then hanging up; he found microphones secreted in his radio set, and his garage mechanic discovered a small sending set under the dashboard of his car; the doors of his car were pried open, the locks broken; cars repeatedly ran back and forth in front of his place, and he was constantly followed. Reich remarked that if he revealed all that had happened to him, people would think he was insane. Was that the purpose, to harass him and wear him down?"* Elsworth Baker, "Wilhelm Reich", *Journal of Orgonomy*, Vol.1, Nos 1&2, 1967, p.51.
80. Reich, Wilhelm: "Emotional Plague" *Selected Writings,* Farrar, Straus & Giroux, 1973, p.467-513; W. Reich, *Character Analysis*, Farrar, Straus & Giroux, 1961, p.504-539. W. Reich, *The Murder of Christ*, Farrar, Straus & Giroux, 1971.

Chapter 5: The *Orgasmatron* Aftermath

1. Turner, Christopher: *Adventures in the Orgasmatron*, Farrar, Straus & Giroux, NY 2011.
2. Report originally posted online the OBRL-News Blog, 13 June 2012:
    http://obrlnews.wordpress.com/2012/06/13/fighting-defamatory-slanders-from-uk-channel-4wag-tv-sex-researchers-program
    http://www.orgonelab.org/Channel4WagTV.htm
3. E-mail to James DeMeo from Ruth Mayer of Wag-TV, 24 Jan 2011.
4. "Kinsey's Paedophiles", Yorkshire TV production for Channel 4, aired 10 Aug. 1998. On YouTube: http://www.youtube.com/watch?v=htAUysRPvNs
5. *The AIDS Catch"* and *"AIDS and Africa"*. Meditel/Channel 4 productions, 1993. Both items are posted to YouTube.com and Vimeo.com. Also see Duesberg, Peter: *Inventing the AIDS Virus*, Regnery 1998, especially Chapter 9 and p.326-327 on the toxicity of AZT. Also see: duesberg.com and virusmyth.com
6. "The Great Global Warming Swindle", Channel 4 production, 2007.
7. DeMeo, James: "Has the AIDS Tragedy been Distorted by Media and Partisan Politics?" Book Review, *Macrocosm USA*, ed. by Sandi Brockway, Macrocosm Publishers, Cambria, CA, 1992, p.130. Fumento, Michael: *The Myth of Heterosexual AIDS*, Regnery 1993.

8. DeMeo, James: "HIV is Not the Cause of AIDS: A Summary of Current Research Findings", *Pulse of the Planet* 4:99-105, 1993.

9. *Open Letter To the Scientific Community* as drafted by *The Group for the Scientific Reappraisal of the HIV-AIDS Hypothesis,* published in *Science,* vol.267 pp.945-946, 17 Feb. 1995.
   http://www.virusmyth.com/aids/data2/letterscience.htm
   http://www.sidasante.com/contacts/group.htm

10. The three scientific petitions were: A) *The Manhattan Declaration on Climate Change: "Global warming is not a global crisis",* drafted by the International Climate Science Coalition, 4 March 2008. B) *An Open Letter to President Obama on Climate Change: "With all due respect Mr. President, that is not true",* drafted by the Cato Institute, 19 Nov. 2008. C) The *U. S. Senate Minority Report: "More Than 700 International Scientists Dissent Over Man-Made Global Warming Claims Scientists Continue to Debunk 'Consensus' in 2008 & 2009",* United States Congress, Washington, DC, 11 Dec. 2008. For weblinks to these documents, see the listings for 2008 at:
   http://www.orgonelab.org/demeopubs.htm

11. DeMeo, James: *"The Suppression of Dissent and Innovative Ideas In Science and Medicine, a Work in Progress."*
   http://www.orgonelab.org/suppression.htm

12. DeMeo, James: *Saharasia: The 4000 BCE Origins of Child-Abuse, Sex-Repression, Warfare and Social Violence, In the Deserts of the Old World,* Natural Energy Works, Ashland, Oregon 1998. (Revised and updated second edition in 2006).

13. DeMeo, James: *The Orgone Accumulator Handbook: Wilhelm Reich's Life-Energy Science and Healing Tools for the 21st Century, With Construction Plans,* Third Revised Edition, Natural Energy Works, Ashland Oregon, 2010.

14. DeMeo, James (Ed.): *Heretic's Notebook: Emotions, Protocells, Ether-Drift and Cosmic Life Energy, With New Research Supporting Wilhelm Reich* (Pulse of the Planet #5) Natural Energy Works, Ashland, Oregon 2002.

15. http://www.youtube.com/user/naturalenergyworks

16. http://www.channel4.com/programmes/the-sex-researchers/4od

17. http://www.ofcom.org.uk/

18. Complaint by Dr James DeMeo, The Sex Researchers, Channel 4."16 *Ofcom Broadcast Bulletin,* #201, 5 March 2012, p.15-30. Internet posted:
   http://stakeholders.ofcom.org.uk/binaries/enforcement/
   broadcast-bulletins/obb2001/obb201.pdf

19. Channel 4 Response to Fairness Complaint by Dr James DeMeo, The Sex Researchers, Episode 2, Channel 4, tx 23 June 2011, Case No. 1-183401588

20. Josie Glausiusz, "Pre-digital dreams", *Nature* 488:270, 16 August 2012.
   http://www.nature.com/nature/journal/v488/n7411/full/488279a.html

21. CSICOP: *Committee for the Scientific Investigation of Claims of the Paranormal.* CSICOP – rebranded today as the *Center for Skeptical Inquiry* (CSI) –is directed and run by non-scientists and does not undertake independent investigations, though they do have an advisory board of scientists who lend their name to the CSICOP/CSI cause. Maddox was a CSICOP/CSI "Fellow", and frequently invited speaker at various "skeptic club" conferences. He

allied with the "skeptics" towards the goal of censoring mainstream media of whatever they deemed to be "pseudo science". For example:
   http://nzsm.webcentre.co.nz/article1419.htm
   http://www.buffalo.edu/ubreporter/archive/vol27/vol27n31/n7.html

22. Schiff, Michel: *The Memory of Water: Homeopathy and the Battle of Ideas in the New Science,* Thorsons Publishers, UK 1998. Garfield, Eugene: "Contrary to Nature?" *The Scientist,* 5 Sept.1988.
   http://www.the-scientist.com/?articles.view/articleNo/9717/title/
   Contrary-to-Nature-/

23. Maddox, John: "News and Views: Duesberg and the New View of HIV," *Nature* 373: 189, 19 January 1995. Also see: Duesberg, Peter & Harvey Bialy: "Duesberg and the Right of Reply (According to Maddox-*Nature),"* in *Genetica* Monograph "AIDS: Virus- or Drug-Induced?" Kluwer Academic Pub, Dordrecht, Netherlands, 1995. http://www.duesberg.com/papers/ch12.html

24. Hodges, Andrew: *Turing,* Routledge 1999, p.53-54. Also see Wansell, Geoffrey: "How Britain Drove its Greatest Genius Alan Turing to Suicide, Just for Being Gay", *Daily Mail,* 11 Sept. 2009.

25. Reich, W.: "On Homosexuality", in *Children of the Future,* Farrar, Straus & Giroux 1983, p.200-202. Hodann, Max: *History of Modern Morals,* Wm. Heinemann, London, 1937, p.57.

26. Reich, Wilhelm: *The Einstein Affair,* Orgone Institute Press, Rangeley, Maine, 1953. see especially AE's letter of 7 Feb.1941 and Reich's response of 20 Feb.1941.

27. anon. *Nature,* Volume 491, Page 191 (08 November 2012)
   http://www.nature.com/nature/journal/v491/n7423/full/491191b.html

28. Gioni, Massimilano and Gary Carrion-Murayari, *Ghosts in the Machine,* New Museum / Skira-Rizzoli, NY 2012.

29. Turner, Christopher: "Thinking Inside the Box", *Cabinet* No.42, Summer 2011, p.13, 15, 19-23.

30. Turner, Christopher: "Wilhelm Reich: The Man who Invented Free Love", *The Guardian,* 8 July 2011. http://www.guardian.co.uk/books/2011/
   jul/08/wilhelm-reich-free-love-orgasmatron

31. http://www.newmuseum.org/exhibitions/view/ghosts-in-the-machine

32. DeMeo, James, et al.: "In Defense of Wilhelm Reich: An Open Response to Nature and the Scientific/Medical Community" , *Water,* 4:72-18, 2012.
   http://www.waterjournal.org/volume-4

33. http://en.wikipedia.org/wiki/Wilhelm_Reich.

34. Sharaf, Myron: *Fury on Earth: A Biography of Wilhelm Reich,* St. Martin's Marek, NY 1983.

35. Online Emotional Plague Bibliography:
   http://www.orgonelab.org/bibliogPLAGUE.htm

Chapter 6: Makavejev's *WR Mysteries*

1. Makavejev, Dusan, Director & Writer: *WR Mysteries of the Organism.,* Criterion Films, Cinema and VHS release in 1971; DVD version in 2007.

Chapter References: 6. *WR Mysteries*

2. Bell, John & Koopman, Barbara: "Film Review: WR-Mysteries of the Organism," *Journal of Orgonomy*, 5(2):227-233, Nov. 1971.
3. Makavejev, Dusan, Director & Writer: *Sweet Movie,* Criterion Films, Cinema and VHS release in 1974; DVD version in 2007.
4. Atkinson, Michael: "Dusan Makavejev, Heavy Petting", *ICF Fix, AMC Network* 2007.
    http://www.ifc.com/fix/2007/06/dusan-makavejev-heavy-petting
5. Holm, D.K.: "Nocturnal Admissions, The Reading Room", *Fred Entertainment,* 2007. http://www.asitecalledfred.com/2007/06/11/nocturnal-admissions-dvd-review-wr-mysteries-of-the-organism-sweet-movie/
6. Rovi, Nathan Southern: *Sweet Movie* Overview, online at multiple websites, such as: http://www.rottentomatoes.com/m/sweet_movie/
7. Muller, Winfried: "Aktionsanalytische Organisation (AAO)", Internet Feb. 2011. http://www.religio.de/therapie/aao/aao.html
8. Makavejev, Dusan: *WR: Mysteries of the Organism. A Cinematic Testament to the Life and Teachings of Wilhelm Reich,* Bard/Avon Books 1972.
9. Blackledge, Catherine: *The Story of V: A Natural History of Female Sexuality,* Rutgers University Press, 2004, p.264.
10. See for example, the letter from Prof. Philip Bennett of 27 Dec. 2008, posted as "Another Smear of Reich: The Story of V."
    http://obrl.blogspot.com/2008/12/another-smear-of-reich-story-of-v.html
11. DeMeo, James: *REVIEW: Outrageous Lies In this Book,* November 27, 2007.
    http://www.amazon.com/Story-Natural-History-Female-Sexuality/dp/0813534550
12. Film Reference, Dusan Makavejev
    http://www.filmreference.com/Directors-Lu-Mi/Makavejev-Du-an.html
13. The File Room: Dusan Makavejev's "W.R.: Mysteries of the Organism
    http://www.thefileroom.org/html/1250.html
14. For example, the 1974 advertising poster for the German release of *Sweet Movie* shows a naked woman holding and licking a huge ice-cream cone bearing the face of Karl Marx. The film itself included a scene of a giant head of Marx planted on the prow of a river barge, upon which various pornographic scenes were filmed.
    http://shenanitims.wordpress.com/2012/05/22/off-season-reviews-sweet-movie-1974/

In Defense of Wilhelm Reich

Additional Materials for Consultation:
Many of these items are internet-posted.
A simple online search should locate them.

Baker, Courtney F. (pseud. Rosenblum, C.F.): "An Analysis of the United States Food and Drug Administration's Scientific Evidence Against Wilhelm Reich, Part II, the Physical Concepts", *Journal of Orgonomy*, 6(2):222-231, 1972.

Baker, Courtney F. (pseud. Rosenblum, C.F.): "An Analysis of the Food and Drug Administration's Scientific Evidence Against Wilhelm Reich, Part III, Physical Evidence", *Journal of Orgonomy*, 7(2):234-245, 1973.

Blasband, Richard: "Book Review: Adventures in the Orgasmatron, How the Sexual Revolution Came to America by Christopher Turner", *Subtle Energies,* 2011.

Blasband, Richard A.: "An Analysis of the United States Food and Drug Administration's Scientific Evidence Against Wilhelm Reich, Part I, the Biomedical Evidence", *Journal of Orgonomy*, 6(2):207-222, 1972.

DeMeo, James: with 23 other signatories: "In Defense of Wilhelm Reich: An Open Response to *Nature* and the Scientific /Medical Community", *Water Journal*, V.4, p.72-81, 2012. Internet Posted: http://www.waterjournal.org/volume-4

DeMeo, James: "Postscript on the Food and Drug Administration's Scientific Evidence Against Wilhelm Reich", *Pulse of the Planet* 1:18-23 Spring 1989.

DeMeo, James: "Anti-Constitutional Activities and Abuse of Police Power by the US Food and Drug Administration and Other Governmental Agencies", *Pulse of the Planet* 4:106-113 1993.

DeMeo, James: "Response to Martin Gardner's Attack on Reich and Orgone Research in the Skeptical Inquirer", *Journal of Orgonomy,* 23(1):126-133, 1989; also in *Pulse of the Planet* 1(1):11-17, 1989; German translation "Irrationale Angriffe Gegen Reich und die Orgonforschung", *Lebensenergie,* 1:24-32, 1990.

DeMeo, James: "The Suppression of Dissent and Innovative Ideas In Science and Medicine: A Work In Progress", First presented to a Special Session at the Annual Meeting of the *Association of American Geographers,* San Francisco 1994; also presented to Special Sessions at the *Southwestern and Rocky Mountain Division, American Association for the Advancement of Science,* at the 72nd Annual Meeting, 1996, Northern AZ Univ., Flagstaff, Arizona, and 74th Annual Meeting, Mesa State College, Grand Jct. Colorado. Internet Posting: http://www.orgonelab.org/suppression.htm

Additional Materials for Consultation

DeMeo, James: "Editor's Postscript: CSICOP, Prometheus Books, Pornography and the *Journal of Pedophilia?!*", *Heretic's Notebook: Pulse of the Planet* 5:65, 2002. (See Appendix, p.217)

DeMeo, James: "Open Letter on Joel Carlinsky and the Organized 'Skeptics' War Against American Orgonomy," 2005. Internet Posted: http://www.orgonelab.org/carlinskyletter.htm

Greenfield, Jerome: *Wilhelm Reich versus the USA,* W.W. Norton, NY 1974.

Greenfield, Jerome: "Wilhelm Reich: 'Alien Enemy'", *J. Orgonomy,* 16(1):91-109, 1982.

Greenfield, Jerome: "Examination of Wilhelm Reich by the Immigration and Naturalization Service", *J. Orgonomy,* 16(2):277-284, 1982.

Greenfield, Jerome: "Reich and the INS: A Specific Plague Reaction", *Journal of Orgonomy,* 17(2):205-226, 1983.

Laska, Bernd: LSR-Projekt: http://www.lsr-projekt.de/wr.html

Martin, Jim: *Wilhelm Reich and the Cold War,* Flatland Books, Mendocino, CA, 2000.

Martin, Jim & Thomas, Kenn: "Toxic Disinformation: Joel Carlinsky's Bonfire of Insanity", *Steamshovel Press.* Internet Posting to: http://www.umsl.edu/~skthoma/carl.htm

Reich, Wilhelm: *Conspiracy: An Emotional Chain Reaction,* Orgone Institute Press, Rangeley, Maine, 1954.

Reich, Wilhelm: *Response To Ignorance,* Wilhelm Reich Biographical Material, History of the Discovery of the Life Energy, documentary supplement No. 1, A-XII-EP, Orgone Institute Press, Maine, 1955.

Reich, Wilhelm: *The Red Thread of a Conspiracy,* Wilhelm Reich Biographical Material, History of the Discovery of the Life Energy, Documentary Supplement No. 2, A-XII-EP, Orgone Institute Press, Maine, 1955.

Reich, Wilhelm: *Atoms for Peace Vs. the HIG (Hooligans In Government),* Wilhelm Reich Biographical Material, History of the Discovery of the Life Energy, Documentary Supplement No. 3, A-XII-EP, Orgone Institute Press, Maine, 1956.

Wilder, John: "CSICOP, Time Magazine, and Wilhelm Reich", *Pulse of the Planet* 5:55-66 2002.

Wolfe, Theodore: *Emotional Plague Versus Orgone Biophysics: The 1947 Campaign,* Orgone Institute Press, NY, 1948.

Wyvell, Lois: *The Jailing of a Great Scientist in the USA,* privately published, NY, 1956; reprinted in Wyvell 1993. Includes "An Eyewitness Report of the Burning of Scientific Books in the USA, 1956", by Victor Sobey.

Also see the "Related Items to Consult on the Skeptics" (p.222)

In Defense of Wilhelm Reich

Major Books By Wilhelm Reich as republished by Farrar Straus & Giroux:
Available from bookstores or online from
Natural Energy Works   http://www.naturalenergyworks.net

The Bioelectrical Investigation of Sexuality and Anxiety
The Bion Experiments: On the Origin of Life
The Cancer Biopathy (Discovery of the Orgone, Volume 2)
Character Analysis
Children of the Future: On the Prevention of Sexual Pathology
Cosmic Superimposition: Man's Orgonotic Roots in Nature
The Early Writings of Wilhelm Reich
Ether, God and Devil
The Function of the Orgasm (Discovery of the Orgone, Volume 1)
Genitality in the Theory and Therapy of Neurosis
The Invasion of Compulsory Sex-Morality
Listen, Little Man!
The Mass Psychology of Fascism
The Murder of Christ (Emotional Plague of Mankind, Volume 2)
People in Trouble (Emotional Plague of Mankind, Volume 1)
Record of a Friendship, Correspondence of
    Wilhelm Reich and A.S. Neill
Reich Speaks of Freud
Selected Writings: An Introduction to Orgonomy
The Sexual Revolution: Towards a Self-Governing Character Structure

Autobiographical/Documentary:
Passion of Youth: An Autobiography, 1897-1922
Beyond Psychology: Letters & Journals, 1934-1939
American Odyssey: Letters & Journals, 1940-1947
Where's the Truth? Letters & Journals, 1948-1957

Special Reports by Wilhelm Reich
Photocopy editions available from the
Wilhelm Reich Museum & Trust   http://www.wilhelmreichtrust.org

The Orgone Energy Accumulator, Its Scientific and Medical Use, Orgone Institute Press, Maine, 1951.

The Oranur Experiment, First Report (1947-1951), Wilhelm Reich Foundation, Maine, 1951.

The Einstein Affair, 1939-1952, Wilhelm Reich Biographical Material, History of the Discovery of the Life Energy, Documentary Volume A-IX-E, Orgone Institute Press, Rangeley, Maine, 1953.

Index

A

abortion 19, 22, 149
aether (see: cosmic ether)
"AIDS Deniers" 182, 191
AIDS hysteria 191
"AIDS tests", flaws, false positive
  reactions 190
Air Technical Intelligence Com-
  mand (ATIC) 79
alcoholism 151
Allen, Woody, movie *Sleeper* with
  "orgasmatron" 118-120, 196
ameba 32, 43
  protoplasmic motions 37
America Anti-Imperialist League
  50
American Stalinists 4
American Association for Medical
  Orgonomy 69
American Assoc. for Advancement
  of Science (AAAS) 99
American Civil Liberties Union
  (ACLU). 49, 50, 52, 86
Am. College of Orgonomy 90
American Conditions and Sexual
  Freedom 151
Am. Journal of Psychiatry 61
American Medical Association
  (AMA) 49, 53
Am. Psychiatric Assocation 82
Am. Psychoanalytic Assoc. 80
American Revolution 154
American Russian Friendship
  Association 125
American sexual reform move-
  ment 151
American Society for Cultural
  Relations with Russia 50
American women status 153
Am. greater sexual, economic and
  political freedom 152
Anderson, William 106

*Annals of the Institute for Orgo-
  nomic Science*, 90
Annals of the Orgone Institute,
  64
Anschluss, Nazis in Austria 42
anti-communism. 9
  (see McCarthy, Joseph)
anti-Slavery movement 153
anxiety versus pleasure 33
apoptosis 43
Archaea 104
arranged marriages 102, 149
astrobiology 43
Austrian Communist Party (KPÖ)
  18, 20
Austrian Social Democrat Workers
  Party (APÖ) 17
Austrian Social Democrats (APÖ)
  19
Austro-Hungarian Empire 13
  invasion of Serbia 13
Azidothymidine (AZT), 190

B

Baker, Courtney, 108
Baker, Elsworth 71, 102, 161
Baldwin, James 145
Baldwin, Roger 49-50, 146
burning of Reich's books (see:
  Reich, bookburning)
Becker, Robert O. 104
Bell, John 92, 202
Benveniste, Jacques
  171, 189, 191
  Nature "hit squad" 189
Berlin Love Parade 149
Berlin Technical Seminar 20
Bernfeld, Siegfried 23
bestiality 162
Big Lies, method of Goebbels and
  the Soviets 172
Big Media "journalism" 175

big-bang creationism 78
bioelectrical investigation of
    sexuality and anxiey 37-39,
    41, 103, 155
Bioenergetics therapy 163
bions-biogenesis (see: Reich bion
    experiments)
biopathies 61
biophoton 211
biophysical armoring 26
bisexuality 135, 142, 144
Black versus Red Fascism. 63
Blackledge, Catherine, slander of
    Reich 206-207
Blasband, Richard 105, 107-
    108, 111, 162
Bletchley Park 192
bluish foggy orgone radiation
    from SAPA bions 42
Boadella, David 89, 113, 120
body-oriented psychotherapy 26
Bolen, Tim, articles on "Skeptic"
    clubs 222
Bolsheviks 14-15
Bookburning by Nazis at
    Opernplatz in Berlin 30
bottled-up emotions 26
Brady, Mildred Edie 34, 51-53,
    57, 64, 68, 73, 91, 124-127, 144,
    146, 158, 164, 200
    Consumer's Reports 88
Brady, Robert
    34, 51, 52, 53, 65, 126
    The Spirit and Structure of
    German Fascism 52
Brady-Gardner slanders 119, 122
Braid, Byron 104
breast cancer 106
Brecht, Bertold 30
Bremmer, Kenneth 107
Brenner, Myron 107
Bullough, Vern 219
Burr, Harold. electrodynamic
    fields 103
Burroughs, William
    118, 119, 120, 137, 139, 144, 146
Bursler, Norman 52, 65
Butterfly Baby-Massage 166

C

California Drought Emergency
    Task Force 99
Cambridge Soviet spy ring.
    20, 52, 64, 126
cancer as a systemic biopathic
    disorder 68
    loss of bioenergy charge 55
cancer mice 4, 40, 45
Carnegie Foundation 49
castes 24
Castro, Fidel 153
Catholic Inquisition 154
Catholic Inquisitors 94
Chamberlain, Neville 42
Channel 4 UK TV:
    AIDS and Africa 181
    The AIDS Catch 181
    The Great Global Warming
    Swindle 181
    Kinsey's Paedophiles 181
character armor and somatic-body
    armor 16
"chembusters" 111
"chemtrails" 213
Chernobyl 182
child abuse 151
child rape & porn 134-135
childhood masturbation as natural
    18
childhood sexual play 24
chiropractic 49, 53
Christian conservatives xiii, 8
Christian Social Action Movement
    50
Churchill, Winston 115
church's stranglehold on family
    and sexual laws. 19
circumcision 160
Civil Rights Movement 153
Clifford, John D., Federal Judge,
    declares "orgone energy does
    not exist" 80
"Climate Deniers" 182
cloudbuster 7, 73, 78, 95-96, 111,
    147, 184
CO2 theory of global warming
    182

Comintern xiii, 30, 34-35, 49, 51-52, 94, 158
Committee for Skeptical Inquiry (see CSICOP)
Committee on Coal and Giant Power 50
Communist Party cooperation with Nazis 48
Comrade Thomas 30
Consumers' Reports 51, 71,158
Consumers' Research 51
Consumers' Union xiv, 51, 54, 68, 73, 158
contempt-of-court 83
contraception 19, 22, 36, 102, 149
coronary occlusion 107
cosmic ether 74, 78, 211
cosmic plasmas 74, 78
Cott, Alan 106, 156, 164
counter-revolutionaries 34
Crombie, Albert 69, 91
cross-cultural studies 24
crystal skulls 211
CSICOP 98, 219, 240

D

Dachau death camp 30
Dante's Inferno 173
dark matter 74, 78, 211
Darwin 167
death-camps 30, 48, 51, 152
DeMeo, James 95-97, 113
  attacked at ISU 98
  attacked by "skeptics" 97
  call to Stand Up for Wilhelm Reich, 194
  Channel 4 complaint 187
  meeting with Matusow in Kansas 238
  protest of Makavejev 203-204
  slander by *Journal of Psychohistory* 93
DeMeo research & writings:
  AIDS criticism 181
  bions 44
  Heretic's Notebook 183
  letter to J. Galassi 176
  oranur/vacor 74

orgone accumulator 58, 60,106, 110, 183, 129
  Reich Cloudbuster 81, 111, 112
  Saharasia 24, 101-102
Democratic revolutions Vs. Divine Power of Kings 154
desertification 24, 101
Deutsch, Arnold, 20, 34
Dew, Robert 104, 107
Dimitrov, Giorgi 34, 51
divorce 19, 22, 149
Dobuans 152
DOR (deadly orgone) 73- 75, 84
blinded and controlled studies 101, 105
drug abuse 151
Duesberg, Peter 181, 189 censored by Maddox & Nature 191
Duranty, Walter, lies for Stalin on Ukranian genocide 25
Durkin, Martin 180
duTeil, Roger 37, 41, 42
Duvall, Albert 162-163

E

East, Jon, film, It Can Be Done xv, 114
Eichel, Edward 132, 219
Einstein, Albert 30, 57-58, 78, 97, 127-129, 167, 192, 194-195
  confirms Reich 128
Eisler, Kurt, Freud Archives 75
Eitingon family, Soviet spies and KGB assassins 50, 119
  Russian-Soviet fur trade 50
electrolytic capacitor 124
Ellis, Havelock 179
emosexuality 222
emotional chain reaction 4
Emotional Plague 93-94, 100, 175, 200
emotional resignation 68
Enabling Act 30
Enigma machine 192
Enlightenment 154
ethics committees 98
EU Constitution 152
"Euro-Superior" theme 151

European conditions 154
exhibitionism 133
extremophiles 104

F

famine trauma 24
Farrar, Straus and Giroux
    publisher 89
FBI investigation of Wilhelm
    Reich 59, 77
Food and Drug Administration
    (FDA) xiv, 5, 53-54, 65-67, 73,
    79-80, 117, 161, 171, 175
    1938 police powers 53-54, 117
    shuts therapeutic natural hot
    springs 54
    investigation of Reich 65
    Injunction against Reich 6, 80,
    161
    fraud, lying to courts 79, 85,
    171
    book-burning 53
    tries to stop republication of
    Reich's books 89
female genital mutilations 150
Fenichel, Otto
    30, 34, 61, 62, 159
    Rundbrief letters 34, 157
    at Menninger Clinic 158
Flexner, Abraham 49
Flexner Medical Education in the
    United States and 49
Flexner Report 49
Flynn, Daniel 92
Folsom Street Fair 149
forced marriages of young girls to
    old men 150
Franco 40, 153
Frankfurt Group 159
Frankfurter, Felix, US Supreme
    Court Justice 49, 50,
    52, 53, 85, 86, 231
    "good friends" with commu
    nist Harold Laski 231
    opposed to US Constitution
    Due Process clauses 232
    lobbied Roosevelt to give a-
    bomb secrets to Stalin 232

"free fuck" 123
freedom peddlers 144
French Academy of Sciences, bion
    discovery presented to 41
French Revolution 154
Freud, Sigmund 13-17, 21, 30, 40,
    46, 147, 179
Freud, Anna 19
Freudo-Marxist psychoanalysts
    61, 157, 159
Fromm, Erich 159
Fuckert, Dorothea 107
Fugs rock band 201
Fukushima 182
Fumento, Michael 181

G

Galassi, Jonathan 176
Galilean prime mover 78
Galileo 94
Gardner, Martin 72, 75, 91-92,
    98, 124, 144, 146
Gauquelin, Michel 171
Geiger-Müller ionization detector
    tubes 7, 110-111
Gell-Mann, Murray 211
genital mutilations 102, 160
genital orgasm 18, 173
    versus the lesser "sexual
    climax" 183
genital versus pregenital im-
    pulses, 16
genital versus pregenital sexual
    activity 141
genitality 9, 15, 36, 117, 141, 152,
    183
German Communist Party (KPD)
    2, 20, 25-26, 54, 147, 160
    denounces Reich 25-26
German Kaiser 14
German munitions factories deep
    in Soviet territory 15
German Psychoanalytic Society
    (DPG) 31, 40, DPG expells
    Jewish members 31, 227
German troops occupy the Rhine-
    land 40
Germany invades Poland 46

Index

Gestapo 35, 147
  bans Reich's books 35
Ginsberg, Alan 118, 119, 120, 137, 139, 144, 146
Glausiusz, Josie 192-193
Goddard, Robert 117, 216
Goethe vii
Gold, Philip 106
Goodman, Paul 144
Göring Institute 31, 40, 227
Göring grabs Prussia 27
Grad, Bernard xv, 104-105
  lukemia mice 108
gravimetric orgonomy 79, 84
Greenfield, Jerome 113
Griffin Bill Committee 52
Group for the Scientific Reappraisal of the HIV Hypothesis of AIDS 181
guillotine 154

H

Hansen, George, analysis of Skeptic clubs 222
Haralick, Robert 95
Harris, Kenyon, Chief of FDA Boston Station 67
Hays, Arthur Garfield 49-50, 68, 85, 86,
  defends communism 51
Hebenstreit, Günter 70, 104, 105
Hefner, Hugh, *Playboy* 135
  support for Kinsey 135
Heine, Heinrich xii
herbalism 49, 53
Herskowitz, Morton xv, 162
Higgins, Mary Boyd 89
High Gods 24
Hinchey, Kevin, *Man's Right to Know* 114
Hindenburg 27, 34
Hirohito, Japan Emperor 153
Hirschfeld, Magnus 142, 179
Hiss, Alger, Soviet spy 51
Hitler, Adolf 2, 15, 30, 34, 42, 153,
  Mein Kampf 17, 31, 228
Hitler-Stalin Pact 30, 46-47, 51, 52

Ho Chi Minh 153
Hodann, Max 30
Holocaust death-camps, role of medical profession 109
Holocaust deniers 182
"holy hand grenades" 213
home-birth 160
homeopathic medicine 49, 53, 189
homoerotic male imagery, of Nazis and Soviets 152
homophobe 150
homosexual bars and nightclub subculture 133-135, 140
  hatred and fear of the opposite sex 149
homosexuality, Reich's view 193-194.
Hoppe, Walter 69, 106, 107
Hoxsey, Harry 53
human cancer 4
hypertensive biopathy 107

I

Illinois State University 97
Immigration and Naturalization Service (INS) 69, 75, 77
impulsive character 15
Industrial-Age CO2 theory of global warming 181
infants and children, abuse of 101
"infectious HIV" hypothesis 181-182, 190
inheritance rules 24
inhibited respiration 26
"intelligentsia" 148, 149
International Communism's 100 million victims 233
International Conference on Orgonomy 116
Int. Journal of Sex-Economy & Orgone Research 59, 163
internet trinket hawkers 212
interstellar medium 78
Int. Psychoanalytic Assoc. 34
  Conference in Lucerne 33. 61, 158
Islamic violence 94, 145, 150, 153
Islamophobe 150

In Defense of Wilhelm Reich

J

Japanese Empire 145, 152-153
  attack on Pearl Harbor 59
John Reed Club 50
Jones, Peter 104
J. Orgonomy 90, 116, 130, 202
J. American Medical Assn. 61, 71
"journalists" as little gods 174
Jung, Carl, Nazi supporter 228

K

Kahn, Albert 86, 238
Kallet, Arthur 51
Kammerer, Paul 14, 171
Karrer, Aurora xiv, 82, 86
Karrer, Aurora, Archive 167
  emotional confusions 166
  on death of Troll 170
  relations with Reich 166
Kautsky, Karl attacks Reich 21
Kavouras, Jorgos 107, 109
Kelly, Charles 89
Kepler 167
Kerouac, Jack 118, 119, 120, 137, 144, 146
Khomeini, Ayatollah 153
King, Jr., Martin Luther 211
King Saud 153
Kinsey, Alfred 9, 103, 116-117, 130-138, 141-143, 156, 160, 164, 179, 183, defined "orgasm" as mere climax 141, advocate of pedophilia 117, 132, 134-139, child sexuality data indicates child-rape 134-138
Kinsey's views clashed with Reich's 136
Kinsey, Hollywood film 135, 137, 138
Knobel, Otto, Reich's assistant on NKVD death-list 37
Koopman, Barbara 202
Kramer, Peter 139, 140
Kreiselwellen spiral waves 78
Kristallnacht 42
Kronstadt sailors 15

KRx Law of Pendulums 84
Kurtz, Paul 92, 219

L

Laboratory Mice 107
Lackner, Stephan 40
Laing, R.D., on Reich 10
Lancet journal 181
Harold Laski 53, 231 dedicates book to Murrow 238
  close friend of Frankfurter 232 (see Frankfurter, Felix), writes Foreword for Brady (see Brady, Robert) 231 f.88
Lassek, Heiko 107
left/right political split 151
leftists different from liberals, similar to Nazis 126
"legitimate" versus "illegitimate" children 19
Leistikow 33
Lenin's secret train ride 14, betrayal of promised democracy 15, secretly works with German militarists 234
Lenin and Trotsky, crimes 233
Leukemia 107
Levine, Emanuel 107
Lewisburg Federal Penitentiary 8, 87
Life and Work of Wilhelm Reich, seminar at NYU 92
life-energy 40
Lindenberg, Elsa 26, 30, 46
Little Ice Age 181, 182
London brothels 135
London Consortium 121
Lowen, Alexander 163
Lysenko 148, 196

M

MacDonald, Helen 75
Maddox, John, *Nature* editor 189
  associate of CSICOP "skeptics" 2, 189
  censors Peter Duesberg 191
  slanders Benveniste (see Benveniste, Jacques)

Maglione, Roberto 111-112
Mailer, Norman 144
Makavejev, Dusan, slander of
  Reich in porn-film *WR Myster-*
  ies of the Organism 91, 201-
  206, included the Muehl sado-
  pedo sex-cult (see Muehl, Otto),
  lied to get film footage 203,
  banned in Yugoslavia 209
male dominance 102
male sexual potency 123
Malignant Melanoma 107
Malinowski, Bronislaw 20-21, 42
man-boy "love" 150
Marcovicz, Digne 114
Marlboro State Hospital 71
Martin, James, *Wilhelm Reich and*
  the Cold War xv, 8, 113, 126
Marx, Karl 18, 147, 167
Marxism xiii, 8, 33, 40, 61,
Marxist death-machine 3, 40, 234
Marxist-Leninist-Stalinist
  agendas 54
Masters & Johnson 156, 179
Masters, Dexter 158
"masturbation therapy", Reich
  falsely accused 62, 155, 173
maternal-infant bond 24, 102
Matthews, J.B., archive at Duke
  University xiv
Matthews, Paul 92
Matusow, Harvey Job 86, 87 (see
  DeMeo, meeting with)
McCarthy, Joseph 51, 86-87
McCarthyites xiii, 8
medical dor-buster 77
medical establishment 159
Medieval Scientism of modern
  medicine 191
Mein Kampf (see Hitler)
Mengele, Josef 138, 148, 196
Menninger Clinic Bulletin 65
Menninger, Karl, slander of Reich
  6, 34, 61, 65, 158
metrosexual 149
midwifery 49, 53
Miller, Henry 123
Mills, Peter 80, 83

Moise, Robert 79
Molly house 137, 162
Mooney-Billings Committee 50
Morris, Robert 111
"Mr. X" (see Kinsey, Alfred)
Muehl, Otto: sex-cult Aktions-
  Analytische Organisation
  (AAO) 205
Mugabe 153
multiple orgasms 134
Munich "peace conference" 42
Murrow, Edward R. 87
Müschenich, Stefan 70, 103
Muslim fanatics 149
Mussolini, Bennito 40, 153
  Pact of Steel with Nazis 46

N

Nagy, Stephen 105
The *Nation* 65, 231
National Institutes of Health
  (NIH) 82, 166
National Popular Government
  League 52
National Science Foundation
  (NSF) 96
Native American herbal remedies
  53
natural heterosexual interests of
  the child 102
natural primary versus secondary
  drives 16, 31
naturopathy 49
National Socialism (NAZIs)
  xiii, 2, 16, 25, 30, 33, 42, 46,
  48, 94, 33, 147, 152, 216
Nazi SA (Sturmabteilung) Brown-
  shirts 21, 27- 28, 30
Nazi-Soviet cooperation 20, 46
Nazi Youth League 30
Neill, A.S., founder of Summerhill
  School 36, 55, 119
  defends Reich 42
neutrino sea 74, 78
New England Journal of Medicine
  181
New Left 160
New Republic magazine 52, 53,

In Defense of Wilhelm Reich

59, 63, 64, 65, 231
New School for Social Research 57, 131
NOAA Weather Modification Office 99
Nordenstrom, Bjorn 104
Normandy invasion 153
Norwegian Communists 75
Norwegian newspapers attack Reich 41
Nunberg, Hermann 158
Nunley, Robert 95-96
Nuremberg Race Laws 35

O

obedience training 24, 26
obstetrics vs. midwifery 102
ocean geothermal vents 43
Oedipus conflict 23, 146
Ofcom, British TV agency 187
Office of Price Administration 52, 126
Offshoots of Orgonomy 164
Ollendorff, Ilse 54, 62, 73, 113, 146, 157, 167
Oppenheimer, Robert 52, 73
optimal orgonotic conditions 110, 242
ORANUR, anti-nuclear radiation effect 64, 73-74
Organic Gardening magazine, FDA tries to ban, burn 53
organism core vs. periphery 16
orgasm reflex 173
orgasm, Reich's 4-phase formula 33
orgasm theory 15
orgastic discharge vs. weaker sexual climax 39, 141
orgastic impotence 142
orgastic potency 16, 18, 127, absent in neuroses and psychoses 18, a primary goal of therapy 18
orgone accumulator 42, 45, 56, 60-61,70, 78, 105-106, 184 abuse of 145

as Faraday cage 4, 42, 56, 78, 124
as hollow capacitor 4, 56, 124
burn healing 71
health benefits 70
not a "cancer cure" 5
not a "sex box" 5
parasympathetic stimulus 70
physical anomalies 55, 58, 73, 111
Orgone Biophysical Research Laboratory (OBRL) 99, 184
orgone energy 4, 32, 42, 45, 55, 61, 78, 110, 160, 211, blue-glowing 78, orgone field meter 62, high-vacuum vacor experiments 73, orgone motor force 7, 64, 67, 71, 79, darkroom scintillations 55, pulsation demonstrator 62
Orgone Flapdoodle nonsense on internet 111, 211-217
orgone therapy 26, 59, 185
"orgonite", "orgonium", "orgone generators", "chembusters", "chemtrails" 211-217
orgonometric equations. 78
orgonomic first aid 160
orgonomic functionalism 79
orgonomic medicine 82
Orgonomy 2, 4
Orgonon, Rangeley, Maine, Reich's home and laboratory 62, 64, 66, 131
origins of life question 3, 104
Orwell, George 115
osteopathy 49, 53

P

Padika: The Journal of Pedophilia 219
parasympathetic versus sympathetic nervous system 39, 103
"Parell, Ernst", Reich's pseudonym 31
patriarchal authoritarian family & culture 16, 24
peaceful societies 151

pedophile attacks on children 102
"pedophile science" 149, 219-222
pedophilia 133, 134, 135, 137,
    142, 145, 161, 219-222
phallic narcissism 123
pharmaceutical industry 159, 175
Philby, Kim 20
pimps 133
pin-up girl 152
Pink, Annie 15
plaster-casters" porn-cult 202
pleomorphism 44
Pol-Pot 153
police shoot striking workers in
    Schattendorf Austria 18
polygamy 150
Pons and Fleishman 148
pregnant and nursing mothers,
    safeguarding 22
premarital taboos 24
Prescott, James 102
prison brothels 152
Professional Patriots 53
Project Bluebook. 83
Prometheus Books 219
prostitution, ending of 19, 22,
    133
protection of children from adult
    seduction 19, 22
Prozac 139
psychiatric drugs 139, 151
psychoanalysis in USSR 19
pulmonary tuberculosis 107
pulsatory functions, expansion
    versus contraction 33
Pulse of the Planet journal 90

Q

quantum metaphysics 78
quantum physics 211

R

radiation hormesis. 73
Rado, Sandor 22, 146, 159
Raknes, Ola 113
Randi, James 189
Rangeley druggist and teens
    harrass Reich 77

Raphael, Chester 71, 75, 107
Red Army 14 (see Trotsky)
Reformation 154
Reich Archive, Harvard Univer-
    sity xiv, 11
Reich, Annie 19, 22, 26, 30, 62
Reich, Eva 79, 86, 107, 160, 164,
    166, 168, 169, 196
Reich, Eva, Butterfly Touch
    Massage 160
Reich, Lore 19, 160, 164
Reich, Peter 62
Reich, Robert 13
Reich versus Kinsey 130
Reich visits the USSR 20
Reich, Wilhelm: Biography
    (see Timeline chapter)
Reich, Wilhelm:
    atoms for peace 83
    attacks after his death 89
    bioelectrical experiments
        32-33, 39-40, 103, 155
    Oslo Laboratory 45
    bion experiments 40-45, 61,
        68, 104-105
    "Bohemians" 144
    bookburning 6, 49, 80, 84-
        85, 88, 173, 227
    "Masturbation Therapy" lies
    155-158, 163, 185
    cancer biopathy 40, 55, 61-
        63, 68, 71, 73, 75, 82,
        106-107, 124
    cancer mice 107
    childhood masturbation
        18, 71
    childhood phobias 20
    children, protection from
        adult seducers 165
    cloudbuster 79-82
    communist conspiracy 171
    confirmations of Reich 100
    Conspiracy volume 82
    Contact With Space 79, 84,
        147, UFOs 77-79
    cosmic superimposition 7,
        72, 78
    "death-instinct" 23

In Defense of Wilhelm Reich

death-lists, Nazi and
    Communist xiv, 186
decriminalization of homo
    sexuality 173
drought, deserts 77, 82
Einstein 57, 77, 128-129
epileptic seizures 21
emotional desert 77
function of the orgasm 1, 4,
    9, 19-20, 33, 40, 127, 141,
    59, 90
Forest Hills, NY 54
Frankfurt Group, Reich not
    a member 159-160
Freud, relations with 75
genitality 16-17, 19
healthy children 71
Kinsey 130 (see Kinsey)
leukemia 107
low-level radiation 73
Marxist ideology 54, 63
Marxist proletariat sexually
    disturbed, just like
    capitalists 159
medical DOR-buster 82
mystical religion 72
Nazi expulsion order 29-30
New School For Social
    Research, 55
opposition to pedophilia
    161, 173
opposition to pornography
    141, 173
oranur 67, 71-72
orgone and ether 69-71
orgone motor force 71
orgone accumulator 72, 116
    cannot provide orgastic
    potency 125
orgone physics 42, 62, 69,
    71, 111
orgonometric equations 73
psychoanalytic changes 16
Reich Blood Test 43, 55
Response to court 6, 80, 83
sand packet bions (SAPA)
    42, 45
sex-economy 24

SexPol platform 21, 36
    (see SexPol)
sexual reform platform 22
sexual revolution 5, 35, 62,
    69, 90
Soviet communism 63, 159
    Red vs. Black Fascism 63
superimposing orgone
    energy streams 72
support for women's rights
    & freedom 173
Sexual Struggle of Youth 23,
    25, 147 (see Romeo &
    Juliet)
therapy method 26, 59, 159,
    185
Tucson experiments 79
University of Vienna 14
work democracy 46, 62, 64
Reich, Wilhelm, defamation and
slanders against (American
Period Only):
    American Thinker 92
    Arch. General Psychiatry 93
    Channel 4/ Wag-TV Sex Res-
    earchers 174-179
    Commentary magazine 91
    Daily Mail 174
    Daily Telegraph 174
    Financial Times 174
    Gestalt Journal 93
    The Guardian 94, 122, 174,
    196
    Herald Scotland 174
    The Independent 174
    Int. Socialism journal 92
    J. Am. Med. Assn.117, 158
    J. History Behav. Sci. 93
    Lingua Franca 92
    London Review of Books 91,
    94, 118-119, 122, 137-138,
    170, 174
    London Times 174
    Nature magazine 93, 174, 179,
    181, 188, correction of
    slander: 195-196
    New Humanist 174
    New Museum 192-193, 196

New Republic 92, 125, 126, 158
New York Observer 174
New York Times 25, 91-94, 174
New York Times Review of Books 92, 174
New Yorker 174
Psychiatry: Interpersonal & Behavioral 93
Psychology Today 93
Psychosomatic Medicine 62
Salon.com 92
San Francisco Chron. 174
Scientific American 75, 93
Skeptic 92
Skeptical Inquirer 92
Slate.com 139, 140, 174
Sunday Telegraph 122, 174
Time mag. 65, 87, 91-92, 143, 144, 147, 148, 196
Wall Street Journal 174
Wikipedia 179, 197-198
WR Mysteries film 201
Reich-Rubinstein, Annie 59, 158, 160, 164, 185
Reich, William, confusion with Wilhelm Reich 59, 69
Reichstag 21,27, 30, 51
Reichstag Fire Decree 27
Reisman, Judith 132, 138, 219
Renaissance 154
Rene Guyon Society 163
Report on injury to Reich's dog Troll 170
resistance analysis 16, 26
respiration 147
respiratory blocking 16, 68
rheumatoid arthritis 107
"right-wing American McCarthyites" xiii
Ritter, Paul and Jean 79, 89
roaring 20s 151
Rockefeller Foundation 49
Rodale, J.I., FDA attacks 53
Romanov family murdered 14
Romeo & Juliet 1, 9, 24, 102, 161
Roosevelt, Eleanor 73
Roosevelt, Franklin D. 23, 51, 52, 53, 54, 73, 126

Rosenbergs, Soviet spies 51
Rubinstein, Arnold 30, 158
Ruppelt, Major Edward J. 83
Rush, Benjamin, on freedom of choice in health care vii
Russian Democratic Soviets 14
Russian Imperial Duma 14
Russian Provisional Government 14, 153
Russian Revolution, 1917 February vs. October 14

S

Sacco & Vanzetti 51
sadomasochism 135, 142
Saharasia 24, 146, 151, 154, 183
Salinger, J.D. 144
San Francisco porn scene 135
Savage, Dan 140-141
Schilder 15
Schjelderup, Harald 32
Science magazine 181
Screw magazine 201
Int. Symposia on Pleomorphic Microorganisms 105
self-regulated vs. compulsive behavior & morality 16, 31
Senf, Bernd 113
sex-economy 26, 31, 183
sex-offenders 133
SexPol 18-20, 22, 30, 36, 40, 103, 122, 141,148-149, 154, 165
Sex Researchers, UK TV 179 (see Channel 4, WagTV)
sexual anesthesia 18
Sexual Behavior in the Human Female (see Kinsey, Alfred)
Sexual Behavior in the Human Male (see Kinsey, Alfred)
sexual climax vs orgasm 141
sexual dysfunction 156
sexual latency 23, 146
sexual narcissism 133
sexual rights of unmarried youth 19
sexual stasis 68
sexually healthy and non-violent society 151

Sharaf, Myron 112, 120, 199
Siersted, Ellen 31, 164
Silvermaster, Nathan and Helen 52, 65, 126
Silvert, Michael 83, 163
  betrayal of Reich 160
  seduces child patient 161
  suicide 88
skeptic clubs 97. 99, 117, 162, 163, 165, 174, 189, 192, 200, 212
  promotion of pedophilia, sexual perversity 164, 219
skin cancer 107
slavery 24
Sobey, Victor 106, 107
Socialist Association for Sex Hygiene and Sexological Research 19
Soviet archives xiv
Soviet Cheka 14
Soviet collusion with German militarists 18
Soviet gulag slave-labor-death camps, 48, 51, 233
Soviet NKVD death-list, Reich's name on xiv, 34, 37-38, 54,
  Reich's assistant Knobel 37
  3000 names reported 34
Soviet KGB 20
Soviet spy-rings 8
Soviet "worker's paradise" 18
Soviet-planned starvation geno-cide in the Ukraine 25, 51, 232
Soviets invade E. Poland 16, 46
Soviets invade Finland 46
Soviets Without Communists 15
Spain is betrayed by Stalin 40, 234
Spain Republican brigades 40
spectrographic UV absorption of orgone-charged water 56
Stalin, Josef 17, 30, 34, 48, 153
Stalinists xiii
Straight, Michael Whitney 20, 52, 64, 126
Straus, Roger 89, 90, 174
Summerhill School 36, 55, 118-119, 146

suppression of women and sexuality 24
Svoboda, Antonin 114, 243
swaddling 102
Sweeney, Judge George C. 83
Swift, Jonathan vii, 179

T

Templeton, Clista 67
Three Mile Island 182
Tolstoy, Leo vii
transvestites 135
Treaty of Versailles 14, 15
Trobriand Islanders 20, 151, 152
Trobriand society 23
Troll hit by a car 170
Tropp, Simeon 106, 107
Trotsky 14, 15, 17, 34, 37-38, 40
  exiled from Russia 17
  Lenin's executioner 224
  "war communism" & terror 16, 17, 224, 232
Trotta & Marer 108
Truman, Harry 126
Turing, Alan 192, 193, 197
Turner, Christopher: Adventures in the Orgasmatron
  attacks Summerhill School 118
  background, credentials 121
  emotional reactions 199
  Grand Inquisitor 172
  omissions 116, 130, 132,
  public support 173
  source of slander for Chan nel 4 TV, Nature mag., Wikipedia 187, 195
  supports Kinsey's "research" & agenda 136
  denied access to Reich Archive 170

U

UFO flight-tracks 147
UFOs 84
Ukranian genocide 25, 51, 232
unhappy marriages 36

Index

Unity Association for Proletarian
Sexual Reform and Protection
of Mothers 21
University of Kansas 81, 95-96,
111
University of Marburg 105
University of Miami 98
University of Oslo 31-32, 39, 103,
116, 155, 185
University of Vienna 15, 104, 105
US Air Force Technical Command
83
US Constitution 6, 76, 80, 86
US State Department 75
US Supreme Court xiv, 7, 49, 83-
84, 159
uterine cancer 107

V

VACOR (Orgone Charged Vacuum)
74
vaginal blood taboos 102
vegetative currents 39
vegetotherapy 26, 41, 59, 185
venereal disease, elimination of
22
Venona decryptions 52
Victor Berger National Foundation 50
Vienna Psychoanalytic Polyclinic
15
Vienna Psychoanalytic Society
(VPS) 15-16
Völkischer Beobachter, Nazi paper,
denounces Reich 27

W

Wagner-Juaregg 15
WagTV Ruth Mayer 179, 180
WagTV Stephen Kemp 180
WagTV Vicky Booth 180
Wall Street Stock Market crash
20
Wallace, Henry 52, 64, 85, 126
Walter, Pierre 221-222
Warburg, Otto 14
Warsaw bombed by Nazis 46

Washington, William 67, 71
Water journal 197
water memory 189
weather and orgone energy 73
Weatherwise magazine 93
Webb, Robert 184
Weimar Germany 142, 153
Wertham, Frederic 63, 125
Western media, Stalin 35
Weverick, N. 107
Wharton, W., Chief, FDA Eastern
District 67
wife-swapping 142
Wilhelm Reich Trust & Museum.
66, 89, 114, 116, 174
burglary of Museum 98
Wilmers, Mary-Kay 119
Wolfe, Theodore 5, 41, 54, 69
women's freedom 19
Wood, Charles A. 67
World League for Sexual Reform
(WLSR) 20
"World Pornography Conf" 220
World Wars I and II 13-14, 152
WR Mysteries of the Organism, 91
(see Makavejev)
Wright brothers 117, 216
Wyvell, Lois 89, 164

Z

Zedong, Mao 153

In Defense of Wilhelm Reich

About the Author

James DeMeo, PhD, formally studied the Earth, Atmospheric, and Environmental Sciences at Florida International University and the University of Kansas, where he earned his PhD in 1986. At KU, he openly undertook graduate-level natural scientific research specifically focused upon Wilhelm Reich's controversial discoveries, subjecting those ideas to rigorous testing, with positive verification of the original findings. DeMeo subsequently undertook field research in the arid American Southwest, Egypt, Israel, sub-Saharan Eritrea, and Namibia, Africa. His work on the *Saharasia* question constituted the most ambitious global cross-cultural research study to date, on the subjects of human behavior, family and sexual life around the world. His published works include dozens of articles and compendiums, and several books, including *Saharasia* and *The Orgone Accumulator Handbook*. He was editor of *On Wilhelm Reich and Orgonomy* and *Heretic's Notebook,* editor of the journal *Pulse of the Planet,* and co-editor for the German-language compendium *Nach Reich: Neue Forschung zur Orgonomie.* DeMeo served on the faculty of Geography at the University of Kansas, Illinois State University, University of Miami and University of Northern Iowa. He is a member of the *American Meteorological Association, Society for Scientific Exploration, Arid Lands Society, Natural Philosophy Alliance, Sigma Xi, International Society for the Comparative Study of Civilizations,* and the *AAAS,* and a former Research Associate of the *American College of Orgonomy.* He is Director of the *Orgone Biophysical Research Lab,* which he founded in 1978. In 1994 DeMeo moved to establish the *Greensprings Center,* a high-altitude research facility in the Siskiyou Mountains near to Ashland, Oregon, exhibiting optimal conditions for sensitive orgone energy experiments. Seminars are periodically offered to serious students. A full list of DeMeo's publications and lectures is given at:

> http://www.orgonelab.org/demeopubs.htm

Or through ResearchGate:

> http://www.researchgate.net/profile/James_DeMeo/

Additional Publications Available from most on-line booksellers, and Natural Energy Works

www.naturalenergyworks.net

SAHARASIA: The 4000 BCE Origins of Child-Abuse, Sex-Repression, Warfare and Social Violence, In the Deserts of the Old World, by James DeMeo

Dr. DeMeo's *magnum opus* on the origins of human violence and biophysical armoring, the first geographical, cross-cultural study of human behavior around the world, using Wilhelm Reich's sex-economic discoveries as a basic starting point, presenting world maps of different behaviors and social institutions. Source-regions (Arabia and Central Asia) for patriarchal authoritarian culture were identified, and migratory-diffusion patterns were traced, back in time, to pinpoint where and how the human tragedy began. Solves the riddle of the origins of human violence and armoring. A breakthrough in the scientific study of human sexuality, psychology and anthropology, and must-reading for every parent, student, professor and clinical worker in the field of human health and behavior. 464 pages, with over 100 maps, photos, and illustrations. Large format with vivid full-color cover, extensive bibliography and index.

Heretic's Notebook: Emotions, Protocells, Ether-Drift and Cosmic Life-Energy, with New Research Supporting Wilhelm Reich, Edited by James DeMeo

Contains 28 insightful essays and research articles by 17 different authors, on natural childbirth, sexuality, archaeology of early human violence, Reich's orgonomic functionalism, exposés on Reich's detractors, Giordano Bruno's work, bion-biogenesis

research, Dayton Miller's ether-drift discoveries, emotional effects in REG (psychokinesis) experiments, new detector for orgone energy, dowsing research, cloudbusting desert-greening experiments in Africa, plant growth stimulation in the orgone accumulator, the orgone energy motor and "free energy", plus UFO research, book reviews, and much more, with color cover photos, text- photos and illustrations. 272 pages

The Orgone Accumulator Handbook: *Wilhelm Reich's Life-Energy Discoveries and Healing Tools for the 21st Century, with Construction Plans,* by James DeMeo

"Orgone Energy Does Not Exist" declared a US federal court judge in 1954, based upon journalist and FDA slanders, ordering the destruction of all publications bearing the forbidden word "orgone". Dr. Reich was later hauled off to prison, where he died. A scandal worse than the "Scopes Monkey Trial" and yet the facts about Dr. Reich continue to be concealed or misrepresented. In this book, former university professor and natural scientist James DeMeo draws from his extensive personal investigations and experiments to present the facts about Reich's discoveries, specifically focusing upon the controversial *Orgone Energy Accumulator,* which is a bona-fide form of *Energy Medicine* used the world over. Here, the reader will learn how to build and safely use orgone blankets and accumulators, with details on orgone-charged *Living Waters* of natural hot springs and healing water spas, and many other related issues. Learn how to concentrate and work with the blue-glowing orgone (life) energy using simple and readily-available materials. A limitless supply is freely available from the atmosphere. Includes many photos and diagrams plus two appended research articles, bibliography, index and weblinked references. **Third Revised and Updated Edition,** 260 pages.

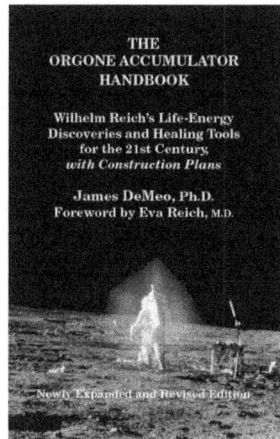

The History of Modern Morals, By Max Hodann, a central participant in the European Weimar-era sexual reform movement. 350+ pages, with a New Introduction by James DeMeo.

www.ingramcontent.com/pod-product-compliance
Lightning Source LLC
Chambersburg PA
CBHW020658270326
41928CB00005B/175

* 9 780980 231670 *